# Engineering Design:
# A Systems Perspective

Bill Fortney
North Carolina State University

Engineering Design: A Systems Perspective

Copyright © 2018 by Bill Fortney

ISBN: 978-1-9874834-1-3

Published by Bill Fortney - 501 Franklin Ave, New Bern, NC 28560.

All rights reserved. No part of this publication may be reproduced, distributed, or transmitted in any form or by any means, including photocopying, recording, or other electronic or mechanical methods, without the prior written permission of the author.

# Dedication

This book is dedicated to God - the Creator of the universe.

> He created with such complexity that we need the analytical thinking of traditional engineering design to understand the details of our world.

> He created with such interconnectedness that we need the systems perspective to understand how it all works together.

To God be the glory!

# Acknowledgments

Thank you to my wonderful and talented wife Pam Fortney. Her labor of love editing my work is what enabled me to create a draft worthy to be reviewed by others.

Listed below are the many people that reviewed this book and contributed to its content.

    Marissa Olberding, Mark Meno, Russell Padgett, Fred Looft, Bob Parys, Dakota Cooper, Jim Wasko, Michelle Pitman, Jim Yankauskas, Jonathan Wiley, Kevin Tierney, Chris Yde, Ramsey Davis, Holly Tucker, Zachary Gupton, Jeff Eischen


Please know that I am very grateful for the investment of your time and that you cared enough to help.

# Contents

Dedication .................................................................................................. iii

Acknowledgments ..................................................................................... iv

Introduction .............................................................................................. ix

**Chapter 1 - Defining Requirements is the First Step to a Successful Design** .. 1

    Reason #1 – To Guide the Design Process ............................................... 3

    Reason #2 – To Satisfy Your Customer's Expectations .............................. 3

    Reason #3 – To be an Effective Designer ................................................. 6

    Summary ................................................................................................ 15

**Chapter 2 - The Systems Perspective is Critical When Designing** ................ 19

    What is a System? .................................................................................. 19

    The Systems Perspective ........................................................................ 27

    Summary ................................................................................................ 28

**Chapter 3 – Engineering Design** ............................................................... 29

    The Typical Stages of Design ................................................................. 35

        Stage One: Requirements Definition .................................................. 35

        Stage Two: Conceptual Design ........................................................... 35

        Stages Three and Four: Preliminary/Detailed Design–Broad View ...... 37

        Stages Three and Four: Preliminary/Detailed Design–Detailed View .... 43

        Stage Five: Fabrication and Testing .................................................... 47

        Stage Six: Implementation ................................................................. 49

        After Implementation ........................................................................ 51

    Design Thinking ..................................................................................... 51

    Summary ................................................................................................ 52

**Chapter 4 - Five Steps You Can Use to Define Requirements** ..................... 55

    General Principles to Follow When Defining Requirements .................... 57

Step 1: Obtain General Understanding of the Problem & Problem Situation .. 59
    Action 1-A: Define Overall NEED ............................................................. 60
    Action 1-B: Define Wider System ................................................................ 62
    Action 1-C: Define Stakeholders ................................................................. 71
    Action 1-D: Define Top-Level Functions .................................................... 72
Step 2: Discover Needs (Wants and Needs) .................................................... 75
    Action 2-B: Explore System to Identify Needs, Interfaces, and Constraints .................................................................................................. 79
Step 3: Clarify Needs ........................................................................................ 88
    Action 3-A: Resolve Questions about Needs .............................................. 89
    Action 3-B: Define Technical Requirements for Operational Needs ...... 89
    Action 3-C: Define Interfaces ..................................................................... 90
Step 4: Express Needs for Customer and Designers ...................................... 91
    Action 4-A: Decide Needs to Include ......................................................... 91
    Action 4-B: Create Requirements ............................................................... 93
Step 5: Validate Requirements ........................................................................ 96
    Action 5-A: Update Documents ................................................................. 96
    Action 5-B: Define Criteria ......................................................................... 96
    Action 5-C: Validate Requirements ............................................................ 97
Summary ............................................................................................................ 98

## Chapter 5 - The Systems Perspective Complements Traditional Design ...... 99
Requirements Definition. ............................................................................... 101
Conceptual Design .......................................................................................... 106
Preliminary / Detailed Design ....................................................................... 109
Summary .......................................................................................................... 114

## Chapter 6 – Performing on a Design Team ...................................................... 115
Attitude is Everything .................................................................................... 116
Define Team Roles ......................................................................................... 117
Resolving Conflict .......................................................................................... 118
Summary .......................................................................................................... 120

**Chapter 7 - Before You Begin Your Design Project ........................................ 121**
    Be Patient ................................................................................................. 121
    Focus on WHAT and Resist Thinking about HOW ............................. 121
    Manage Your Personal Interactions ..................................................... 122
    Learn to Listen ....................................................................................... 122
    Explore Needs as Closely to the User Level as Possible ..................... 122
    Know When to Let Go ........................................................................... 123
    Know When to Quit .............................................................................. 123
    Closing .................................................................................................... 124

**About the Author ............................................................................................ 127**

# Introduction

You have been asked to "design" something for someone. Creating a design seems like a simple enough task, but many who have gone down this road before you will agree that there are serious pitfalls to be avoided. A real story from a friend of mine illustrates this point. My friend Bob worked for a large technology company. The president of the company requested the creation of a special report which required integrating data from several computer systems in real time. Bob was assigned to design a system to create the report, and he thought that this was his big chance. He met with the president to understand what he wanted and then went off and designed a system to create the report. Bob was so proud of his design. He showed the report to the president and waited for the praise to flow. The president reviewed the report for a long time and then said, "This is exactly what I asked you for, but this is not what I need."

Bob learned a valuable lesson about engineering design that day. He learned that engineering design is not just about doing technical work to give customers what they request. He learned that the customer ultimately wants their need satisfied, but often they do not fully understand their need. After his first attempt, Bob worked to understand the president's need and provided a solution that made the president very happy. Bob went on to have a very successful career and satisfied the needs of many customers.

The story of Bob and the lesson he learned about engineering design is not unique. Every successful engineer has learned this lesson and this book is designed to help you learn it as well.

Engineering design is a complex endeavor with many stages, so it is unrealistic for one book to deal in detail with each stage of design. This book will focus on the beginning. It will help you develop a systems perspective of design and show you how to develop requirements for your specific problem. This systems perspective and the requirements you develop at the beginning will serve as guides during each stage of your design project. They will help you avoid the mistake Bob made.

This book takes you on a journey.

> Chapters 1 and 2 lay a foundation. They introduce you to your role as an engineer, the discipline of WHAT before HOW, formal requirements

definition, and the systems perspective. These two chapters present essential concepts for effective design, so read them carefully.

With this foundation, you gain an overview of engineering design and then walk through each stage of the Engineering Design Process in Chapter 3. This presentation of design is structured to give you a solid understanding of what you must accomplish at each stage of design. To be an effective designer, learn the intent of each stage of design and ensure you satisfy this intent for every design you create.

Properly defining requirements is the key to any successful design, so Chapter 4 provides a five-step process you can use to define requirements for your design problem. Review this chapter at the start of each design project, and it will help you avoid the mistake Bob made.

The material from Chapters 1 through 4 are brought together in Chapter 5 through a design example. You are placed on a design team and will see the systems perspective and your traditional engineering design skills working together throughout the design process.

Engineering design is typically not performed alone, so Chapter 6 gives you some tips for working with a design team. If you are like many engineers, you may find working on a design team an extremely difficult task. It is easy to divide work up among people to accomplish assigned tasks, but it takes work to utilize the unique strengths of each person on the team to accomplish a specific task better than any single person on the team could complete it alone. Apply the tips from Chapter 6 to help your team become a high performing design team.

Your journey closes in Chapter 7 with some practical advice to guide you through your design project.

As you travel through this book, please keep in mind that its overall purpose is to improve your effectiveness as an engineer and designer. It is not a textbook on design, systems, or systems engineering, and some terms are used loosely when trying to illustrate various concepts. I am not trying to redefine these terms. I am trying to 1) help you understand key concepts vital to your success as an engineer, and 2) provide you with tools that you can apply right now.

It is very likely that you are presently involved in a design project. The best way to learn is by doing, so take what you read and apply it to your current project.

# Chapter 1 - Defining Requirements is the First Step to a Successful Design

*The customer expects you to help determine everything necessary to satisfy their NEED, and then satisfy it. Determining everything necessary to satisfy the NEED is the essence of defining requirements.*

---

What are requirements and why are they necessary for a successful design? The term "requirements" is used in many different ways. Sometimes it is used to refer to a legally binding document that defines precise characteristics that must be satisfied by your design. It is also used to refer to a list of statements that define what your design must accomplish. Before I define my use of the term "requirements," let's take a step back and consider what is expected out of you as an engineer. Figure 1 illustrates this expectation.

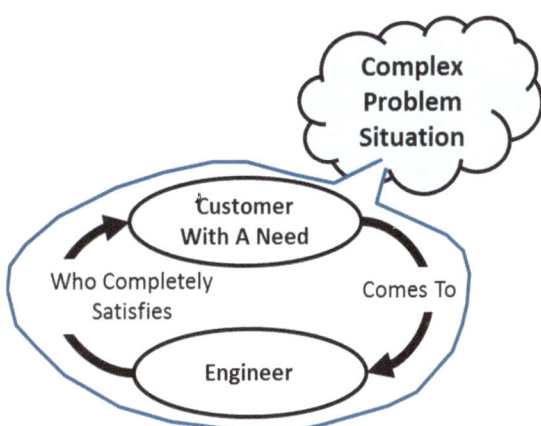

Figure 1: Expectations of an Engineer

Real-world problems occur within complex ("Messy") situations. There are always many technical and non-technical factors that create the problems while simultaneously hiding what is necessary to solve them. Most of the time, the customers themselves do not fully understand the problem or know exactly what their need is.

Customers, with a vague problem within the complex environment, come to you with a need. Please note that I will always refer to this high-level need as "NEED." The customer wants you to satisfy their NEED completely. They do not want you to design for them a specific product, even though they will likely ask for one.

What the customer does want you to do is

- study their complex problem situation,
- accurately understand their NEED,
- determine everything necessary to satisfy their NEED, and
- develop a product (system) that when implemented will satisfy their NEED.

Stop right now and think about this expectation. From the customer's perspective, you are not a "Designer." You are a "Needs Satisfier." You do not just design products that meet specifications or give the customer what they request. You satisfy a customer's NEED. As an engineer, the customer expects you to help determine everything necessary to satisfy the NEED, and then satisfy it. Determining everything necessary to satisfy the NEED is the essence of defining requirements.

The story about Bob from the Introduction illustrates this point. The president of the company had a NEED, but he did not completely understand what it was. He thought that a specific product would satisfy his NEED, so he asked Bob to create the product. Bob created the requested product only to find that it did not satisfy the president's NEED. Bob's design was not a success because it did not satisfy his customer's NEED.

Satisfying the customer's NEED IS engineering, and defining requirements is understanding and defining everything necessary to satisfy the customer's NEED.

When defining requirements, you will discover *wants* and *needs*. A need is something essential to meeting the NEED. A want is something that may be helpful, but technically the NEED can be satisfied without it. On paper, it is easy to make a clear distinction between wants and needs, but it is not so easy when working in the complex problem situation. Often, it is hard to distinguish between a need and a strong want, and sometimes a customer will not be satisfied unless they receive a want. Since the customer will not be satisfied without the want, does this make the want a need? We can discuss this question for a very long time without reaching a definitive answer, so I take the following approach in this book.

- When exploring what is necessary to satisfy the customer's NEED, you are **always** identifying wants and needs, even though you may not always be sure which is which.

- Since you are always looking for wants and needs, I use terms like "needed," "needs," and "necessary" loosely in this book to include wants and needs. When you see these words in the context of defining requirements, remember that I am referring to wants and needs. For example, when you see phrases like below, I am always referring to wants and needs.

    "Define everything necessary to satisfy the NEED."

    "Determine all of the needs."

    "Determine what is needed."

Once all of the needs (remember, I mean wants and needs) are understood, they are reviewed to determine the specific needs (wants and needs) to be addressed by your design. These needs are collected and refined to form the requirements. In this book, requirements are the specific wants and needs that you and the customer agree will be addressed by your system to satisfy their NEED.

> Requirements are the specific wants and needs that you and the customer agree will be addressed by your system to satisfy their NEED.

Defining requirements is critical to a successful design for three key reasons.

## Reason #1 – To Guide the Design Process

At its core, designing is creating something for a purpose - even if that purpose is just to be aesthetically pleasing. Without some understanding of what is needed for your creation (requirements), you as a designer have nowhere to start. Without some statement of what is needed (requirements), you as a designer have no way to judge if one decision is better than another. Without a complete understanding of what is necessary to satisfy the customer's NEED (requirements), you as a designer have no way to know when your design is complete. Defining requirements is critical to a successful design because they give you a clear and complete picture of what you have to accomplish through your design.

## Reason #2 – To Satisfy Your Customer's Expectations

Without defining requirements, you cannot meet the expectations of your customer. Remember that the customer expects you to satisfy their NEED. Figure 2 illustrates the essence of how you accomplish this when designing a new product or system.

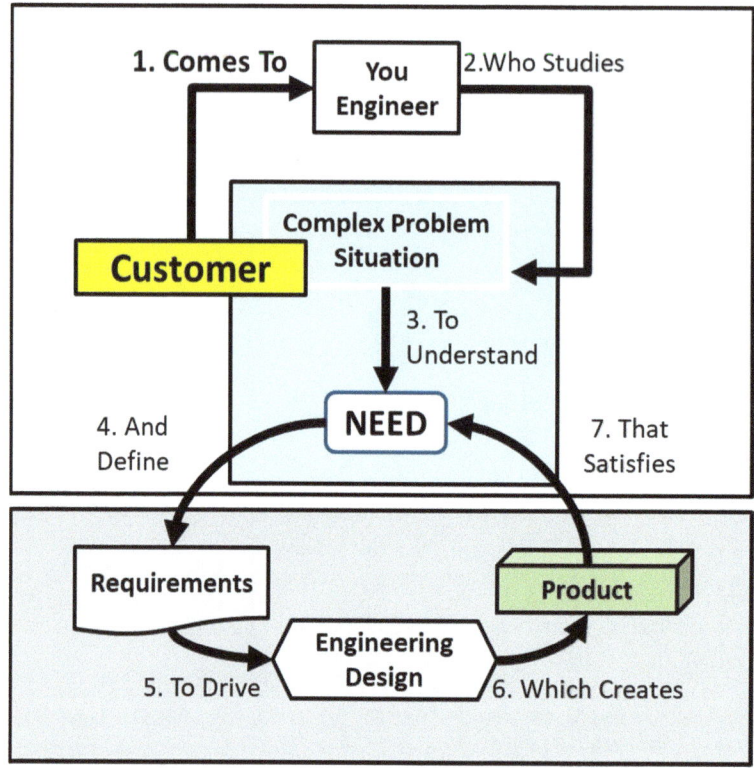

Figure 2: Satisfying a Customer's NEED

Notice the steps for satisfying the customer's NEED from Figure 2.

1. The customer, with a NEED within some complex problem situation, comes to you, the engineer (#1).

2. You study the complex problem situation (#2) to understand the NEED (#3) and to define requirements that fully embody the NEED (#4).

3. These requirements drive your design effort (#5), which creates a product (#6).

4. When implemented within the complex problem situation, your product satisfies the customer's NEED (#7).

You may think of engineering design simply as the bottom three blocks in Figure 2 (given requirements, you design and build a product), but it is not. Like the example with my friend Bob, most of the time the customer does not clearly know what their NEED is. They may understand their NEED from their perspective, but typically they do not understand all of the perspectives which exist within the complex problem situation. If all you do is perform the bottom three tasks based on what the customer tells you, it is very unlikely you will satisfy their NEED. Like Bob, you may give them what they asked for but not what they NEED. The customer depends on your help to define their NEED by performing the tasks in the top portion of Figure 2. They want you to step into the complex problem situation and 1) understand the true NEED, and 2) understand everything necessary to make your design completely satisfy the NEED. They want you to define requirements.

Notice the statement above "understand everything necessary to make your design completely satisfy the NEED." Do not miss the word "everything." Everything means **everything**. Of course, you must understand what the customer wants from your product, but understanding customer expectations is just the beginning of understanding needs. Problems are always multifaceted, and your solution must consider all aspects to be successful. The complex problem situation will generate many needs that you must be aware of as you design.

Imagine that you have been hired to design a piece of test equipment. Your design must be able to perform all required tests, but consider a few failures that could occur by not considering other aspects of the problem situation.

- Your test equipment must be periodically calibrated to satisfy industry regulations, and the company does all calibration through an outside firm that does not come on site. You did not know this, so your design is too large to ship, and the component which requires calibration is not removable. The customer is not satisfied, and identifying the need for off-site calibration up front could have prevented this from happening.

- Your design includes a motor that requires three-phase power. When you begin to install your machine, you discover that the plant does not have three-phase power available anywhere near the machine. The customer is not satisfied, and identifying power needs for your design up-front could have prevented this from happening.

- Your device utilizes brand X smart-electronics. All other equipment in the plant uses brand Y smart-electronics. Your use of brand X means more training for maintenance and extra cost due to the additional spare parts that will need to be stocked. The customer is not satisfied.

Defining requirements is understanding **everything** required to satisfy the customer's NEED. If you do not understand what is necessary to meet the customer's true operational NEED, then it is very unlikely the completed system will satisfy it. If you are not thorough in exploring all aspects of the customer's NEED, then your system may do some things well but be very ineffective at doing other things. A good design is one that makes all users feel like the system was designed just for them. A properly designed system "fits like a glove" when implemented, and properly defining requirements is the key for this to happen.

**Reason #3 – To be an Effective Designer**

As a designer, defining requirements keep you from jumping to a solution too early in the design process. If you are like most engineers, this idea of spending a considerable amount of effort up front understanding and defining the NEED does not sit very well. It is likely that you will "see" the final design solution about halfway through the customer's initial description of the problem to you. Your solution will make complete sense to you, and you know that it will perfectly satisfy the customer. In your mind, all activity between this point and when you get to start designing your conceived solution (like defining requirements) is just a waste of time. It may come as a shock to you, but you are not smart enough to intuitively know everything necessary for a good solution, much less to have the best solution at this point. No one is. Real-world problems are too complex for this simplistic thinking, and moving to a solution too early can greatly limit you and hurt your final design. There is a long list of engineers like Bob who jumped to a solution too quickly and then delivered a product that did not satisfy the customer's true NEED. They delivered a very good solution for the wrong problem.

To be an effective designer, you must learn not to jump to a solution too quickly. You have to learn to address the WHAT before addressing the HOW. I call this "the discipline of WHAT before HOW." Once mastered, this discipline will influence every aspect of your engineering and will be essential to your effectiveness. You will use it for every task you perform as an engineer.

It takes a conscious choice to learn the discipline of WHAT before HOW, and it takes practice to master it. I am introducing you to this discipline now because 1) it is something you must develop to be an effective designer, and 2) it relates closely to defining requirements. During engineering design, properly defining requirements force you to practice the discipline of WHAT before HOW. I will talk more about this later, but first, let me help you better understand the discipline of WHAT before HOW and show you why it is critical for you to develop this discipline.

When faced with a problem, our natural response as an engineer is to start determining HOW to solve it. We jump right to the details such as what the design will look like, how we will power it, and what material we will use. Focusing on these details is starting in the wrong place. Our first response should be to understand WHAT we are truly trying to accomplish. We should begin by asking questions such as "What are we trying to accomplish?" and "What outcome do we want to achieve?" Understanding WHAT is like getting our compass lined up in the right direction and then the HOW takes us there. If we jump straight to the HOW, we will likely end up in the wrong place, and the customer will not be satisfied. Like Bob, we may give them what they asked for, but we will not satisfy their NEED.

When faced with a problem or decision to make, the discipline of WHAT before HOW leads you to a solution using the steps illustrated in Figure 3.

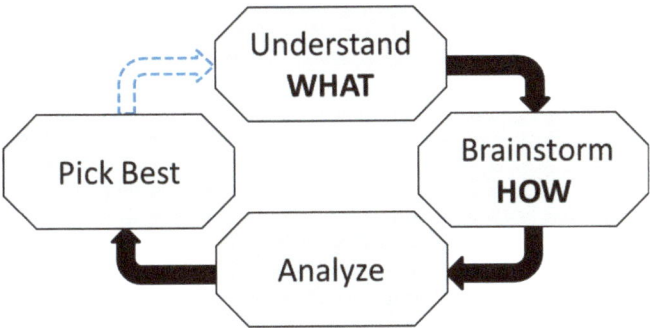

Figure 3: The Discipline of WHAT before HOW

- First, you define WHAT. You clearly understand what you are trying to accomplish and the outcome you want to achieve.

- Next, you brainstorm on HOW to accomplish the WHAT. There are always many HOWs for a given WHAT, and considering these many options is what leads to effective and innovative designs. Most design breakthroughs come while thinking creatively about WHAT.

- Finally, the various HOWs are analyzed, and the one that best fits all of the needs of the situation is selected.

- Determining the HOW typically leads to another WHAT, and the cycle continue.

Effective designers practice the discipline of WHAT before HOW, and it forms the foundation for making engineering decisions at every level of design. For example, it is used at the top level to select a wheeled vehicle over other alternatives to solve some transportation problem. It is then used for every decision made concerning the vehicle down to what type of tires the design utilizes. For some problems, you use the discipline very formally, and for others, the steps in Figure 3 are all done quickly in your head. Regardless, good engineering decisions are made by utilizing the discipline of WHAT before HOW.

Consider the following non-engineering example that illustrates the discipline of WHAT before HOW.

You walk into the break room at your university, and your friend is frantic. He is trying to get a package of crackers from the vending machine, but his selection is stuck and will not fall. He has shaken the machine vigorously to no avail. He is now in the middle of devising an elaborate plan to retrieve the crackers by reaching up into the machine with a quickly designed custom device (your friend is an engineering student).

As someone seeking to learn the discipline of WHAT before HOW, you see this as an opportunity to practice and have the following conversation with your friend.

You:
"What are you doing?"

Friend:
He tells you the story above and shows you his creative device to get the crackers.

You (trying hard not to get sucked in, grab the device, and try it out):
"What are you trying to accomplish?"

Friend (looking at you like you are stupid):
"Get my crackers."

You:
"Why?"

Friend (frustrated with your questions):
"Because I am hungry and need to eat something before my exam in 30 minutes."

You (excited that you finally see the true WHAT for the situation):
"Oh! So what you are trying to accomplish is to eat something and be prepared for your exam. "

Friend (looking at you with a puzzled look):
"Yes, I guess you are right. That is what I need."

You (with a look of victory):
"Well, there are many ways to accomplish getting ready for your exam other than getting your crackers from the machine. You can use another machine, you can walk next door to the market, or you can ask me for some of my lunch. Let's pick the best option and get you ready for your exam."

Yes, this is a silly example, but it illustrates what happens to us so easily in so many different contexts. Notice a few key points from this situation.

- When the problem first occurs, your friend jumps to a HOW – get the crackers out - and defines it as his WHAT.

  > Stop and think about this situation for a minute. It is so easy to believe that the WHAT for the situation is to get the crackers out. The crackers are stuck, and they need to come out. Right? What is it that your friend needs to accomplish above everything else? He needs to be ready for the exam. Getting ready for his exam is his WHAT. Eating something may be part of his true WHAT, but getting the stuck crackers certainly is not. Getting crackers out is a HOW.

- His focus on the wrong WHAT of getting the crackers out is distracting him from his true WHAT (get ready for the exam).

  > Getting the crackers out is just one way to get ready for his exam, but notice that he is not considering any other ways. His focus is on the WHAT of "get the crackers out" instead of "get ready for my exam." All of his activity seems to make sense since in his mind he must get the crackers out. In reality, he does not have to get the crackers out. He does have to get prepared for his exam, but trying to get the crackers out is keeping him from doing this.

- Clarifying the true WHAT changes everything.

  > When you look at the situation in light of getting ready for the exam, it changes everything. Instantly spending time making something to reach inside of the machine to get crackers out does not seem like such a good idea. With a clear picture of the true WHAT (get ready for the exam), it is obvious that there are many other ways (HOWs) to achieve the WHAT. The solution space instantly goes from one option to many options.

- Your friend (the customer) needs your help to understand his true NEED.

  > Like the customers you will work with, your friend is in the middle of a complex problem situation, and he needs you to help him understand his true NEED. The problems you will face as a designer will be much more complicated and will have many more dimensions than this situation, but the customer needs you to bring clarity just like you did for your friend in this simple example. Defining requirements allows you to bring this clarity.

- Your friend (the customer) ultimately must define the WHAT.
  > You help your friend see the true WHAT for the situation, but it is still his choice to accept it. He can listen to you and still decide that he wants to spend his time getting the crackers out.

You may look at this example and think that you would never be as blind as the friend, but engineers do it every day. It is very tempting to skip the WHAT step and jump right to the HOW, but doing so can easily end in disaster. Remember Bob? Bob skipped WHAT and focused only on HOW. Skipping WHAT caused him to develop a very good solution for the wrong problem, and this left the customer's NEED unsatisfied. Skipping WHAT most always results in missing the customer's NEED.

Consider a few simple examples from class design projects that illustrate the importance of the discipline of WHAT before HOW when performing engineering design.

Example #1: Robot ARM

Figure 4 shows the robot arm challenge. Using basic building components and syringes, students must design and build a robotic arm that in the shortest amount of time will

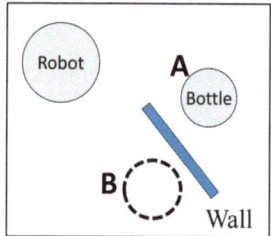

Figure 4: Robot Arm Challenge

- take a 16oz soda bottle from point A,
- transport bottle over the wall,
- set bottle down on circle B,
- release bottle, and
- move the device away from the bottle.

Students typically assign a team member to design and build the gripper, and they usually come up with something similar to the example shown in Figure 5.

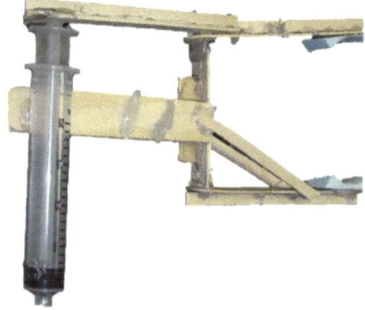

Figure 5: Typical Gripper

Notice the mistake they made. The students defined their WHAT as "design a gripper," but this is a HOW instead of a WHAT. The true WHAT is "hold the bottle for pickup," and using a gripper is only one of the many ways to do this. Consider some other ways to hold the bottle for pickup.

- Two sticks could slide under either side of the lip at the top of the bottle
- A suction cup could attach to the top of the bottle
- A lasso could snag the bottle
- A very sharp prong could pierce the bottle

By skipping the WHAT phase, the students passed over all possible solutions for holding the bottle except one – a gripper. No matter how smart they are, all they will ever develop is a gripper. With no evaluation, they eliminated all possibilities and narrowed their focus to one alternative. This narrowing is what skipping WHAT always does. When you jump straight to HOW, you bypass the analysis of options that lead to the best solution, and doing so too early can be very costly.

Another mistake students make on this project is to define their WHAT as "build a robotic arm." They typically come up with something similar to the one shown in Figure 6.

**Figure 6: Typical Robotic Arm**

One group took to heart the discipline of WHAT before HOW. They correctly defined their WHAT as "get the bottle over the wall and lined up on the target as quickly as possible." They then brainstormed ways to accomplish this WHAT. One idea which received a few laughs was to throw the bottle over the wall. During brainstorming, they went back to the "throw the bottle over the wall" idea, and this discussion eventually became the final design shown in Figure 7.

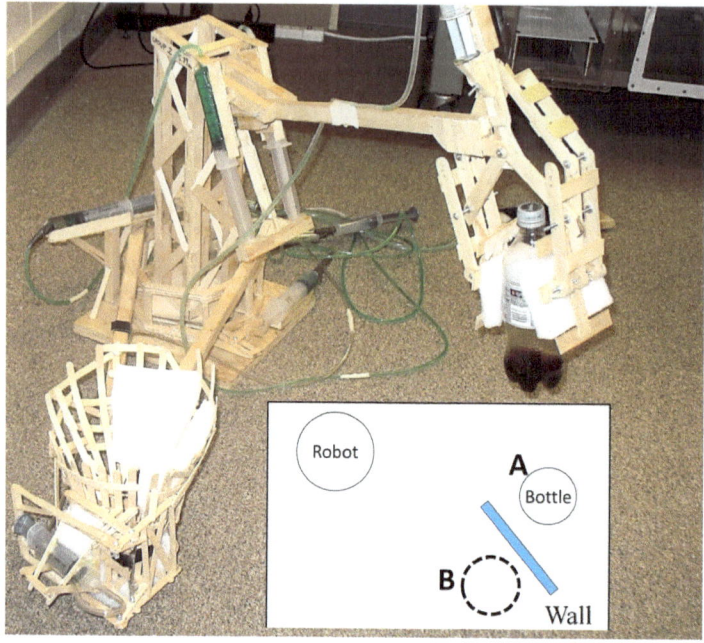

Figure 7: WHAT Driven Robot Arm Design

The design has two independently controlled arms with one arm on each side of the wall. It operates as illustrated in Figure 8.

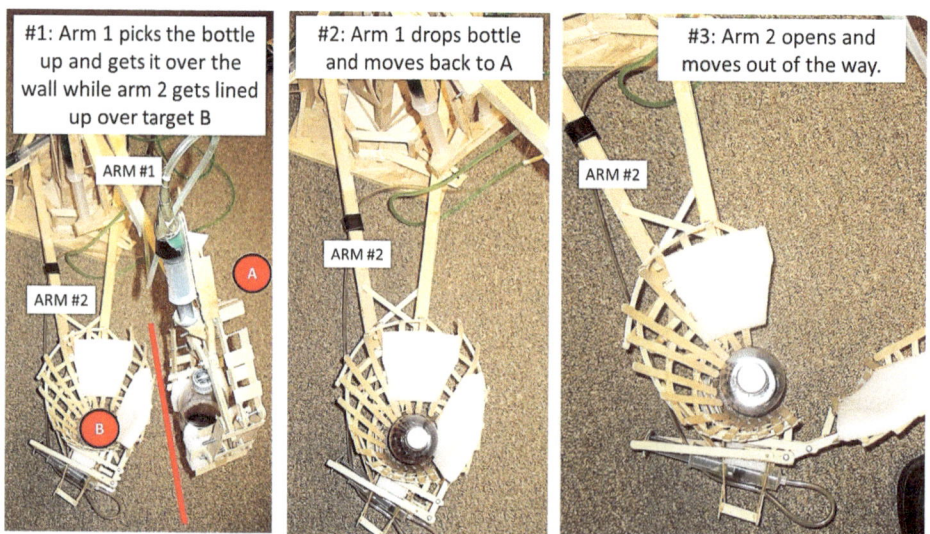

Figure 8: Robot Arm Operation

Arm 1 picks up the bottle and gets it over the wall and roughly over point B. At the same time (parallel tasks instead of sequential), Arm 2 aligns itself perfectly with the target B. Arm 1 drops the bottle and moves. Arm 2 catches the "thrown" bottle, and the bottle slides to the ground perfectly aligned with target B. Arm 2 opens and moves.

This team completed the task in approximately 4 seconds. The closest competitor was over three times longer at 13 seconds.

The design blew everyone away, and notice where they won. They won during the WHAT phase. All of the other teams defined their WHAT as "build a robotic arm," and that is exactly what they created. They completely ignored the first three steps in Figure 3 and focused on one HOW – robotic arm. They cheated themselves by blindly ignoring all possible solutions except one. This is what happens when you do not practice the discipline of WHAT before HOW and jump straight to HOW.

In contrast to the other teams, the winning group applied the discipline of WHAT before HOW. They took a step back and first focused on defining WHAT they were trying to accomplish - "get the bottle over the wall and lined up on the target as quickly as possible." Instead of locking into one narrow idea, they considered many possibilities for achieving the WHAT. An innovative solution emerged for the WHAT. This is what often happens when you take the time to exercise the discipline of WHAT before HOW.

Example #2: Robot Runner

For the Robot Runner competition, students must design and build a vehicle to maneuver as rapidly as possible the maze shown in Figure 9. The exact dimensions are not given, but the maze will have a similar shape to the one shown.

Most students define the WHAT as "build a robot to maneuver the maze" and create an intelligent device that in some way senses the sides of the maze, turns, and makes its way through the maze.

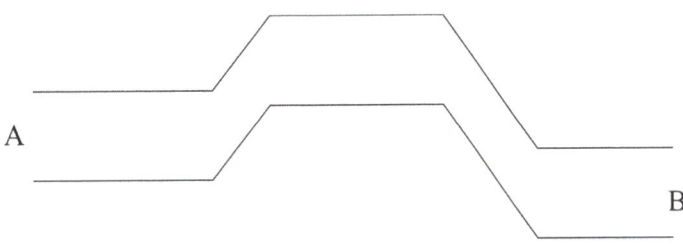

Figure 9: Robot Runner Example

One group applied the discipline of WHAT before HOW and defined the WHAT as "build a vehicle which gets from A to B as quickly as possible." They brainstormed many options and came up with the design shown in Figure 10.

Figure 10: WHAT Driven Robot Runner

The vehicle contained a strong motor and the black circle on the front is a disk that rotates. It forced its way through the maze in a matter of seconds. In fact, it completed the course before all other competitors even made the first course adjustment to their vehicle.

Once again, the team blew all of the competition away because they took the time to consider WHAT, and doing so enabled them to consider all solution approaches instead of just one approach.

If you want to be a good engineer, then develop the discipline of WHAT before HOW and let it drive all that you do. When performing engineering design, defining requirements will help you to implement the discipline of WHAT before HOW. Notice in the examples above that the name of the competition greatly influenced the approach most students took. I never said that the vehicle to maneuver the maze had to be a robot, but most designed a robot. The event was called "Robot Runner," so they designed a robot. Just like I did with my students, the way your customer expresses what they want from you will naturally steer you in a specific direction. This direction may or may not reflect the true NEED or result in an effective solution. The discipline of defining requirements at the beginning of your project is how you avoid the natural pull to follow the customer's direction and jump straight to HOW. Like the example with my friend Bob, most of the time the customer does not fully know what their NEED is. They need you to help them define the WHAT (their NEED), and you do this by properly defining requirements.

Defining requirements is essentially determining WHAT. When you define requirements, you do not determine HOW you will satisfy the customer's NEED. You determine WHAT is necessary to satisfy the NEED. Defining requirements at the beginning of your design process will keep you from jumping to a solution too quickly. Consider the following personal example.

My first job out of graduate school was with a large cabinet company. I was hired to automate one of their wood processing facilities, and my boss made it very clear that my mission was to automate the facility successfully. I could have jumped in and started working on all of the details for HOW to automate the various processes. Instead, I started off by defining requirements. I learned that the plant received orders for parts such as doors and drawer fronts from several cabinet assembly plants. Rough lumber was purchased from lumber mills, and the plant processed it to produce the cabinet parts shipped to the assembly plants.

As I explored needs, I learned that the plant was facing critical changes in both of its key inputs (lumber and orders). Lumber prices were rapidly increasing, and the quality of the incoming lumber was simultaneously decreasing. At the same time, the assembly plants were ordering more frequently and in smaller batch sizes.

It became clear that true NEED was not an automated facility as requested. The true NEED was to maintain the cost competitiveness of the plant in spite of the changes with lumber and orders. With a clear understanding of WHAT, I continued to define the needs and then express them in requirements. We designed and implemented an automated facility which did, in fact, satisfy the true NEED. The plant provided the flexibility necessary to accommodate orders from the assembly plants and maintained cost competitiveness in spite of the rising lumber prices and decreasing quality.

If I had only done what I was asked, "automate the plant," I would not have done it in a way that would have adequately addressed the lumber and order issues. I would have given the customer (my boss) what he asked for, but I certainly would not have satisfied his NEED and, the project would have been a failure.

Effectively defining requirements at the start of the project helped me to exercise the discipline of WHAT before HOW and saved the day as well as my job!

**Summary**

This chapter opened with the question "what are requirements and why are they necessary for a successful design?" Requirements are the specific wants and needs that you and the customer agree will be addressed by your system to satisfy their NEED. Hopefully, you now see the importance of defining requirements and are committed to learning how to define requirements effectively. In summary, requirements are necessary for the following three reasons.

- Requirements guide the design process. At its core, designing is creating something for a purpose, and the purpose comes from requirements. Requirements give you a clear and complete picture of what you have to accomplish through your design.

- Requirements allow you to satisfy the expectations of your customer. Effective design is satisfying the customer's true NEED, and you can only do this by clearly defining requirements. You cannot create a design that satisfies the customer's NEED until you understand and can express the NEED. You understand and express the NEED by defining requirements.

- Defining requirements keeps you from jumping to a solution too early. Defining requirements helps you resist the natural pull of your mind to assume the WHAT and jump straight to HOW. Your customer needs your help to understand their true NEED, and defining requirements allows you to do this.

This chapter also introduced two other ideas critical to effective design - the discipline of WHAT before HOW and an understanding of your role during the design process. These principles are summarized for you in Figure 11 and will be used throughout the rest of this book. Review this figure often during your design project and let these principles guide your thinking during all stages of design.

# Effective Designers...

1) Achieve the Ultimate Goal of Satisfying the Customer's NEED

2) By Using WHAT/HOW Thinking

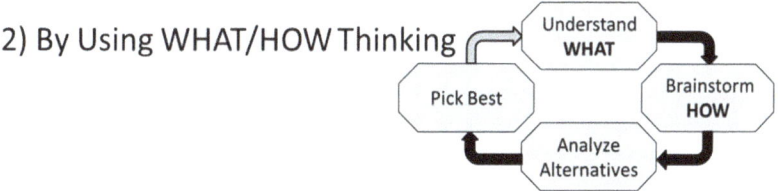

3) During the Entire Design Process

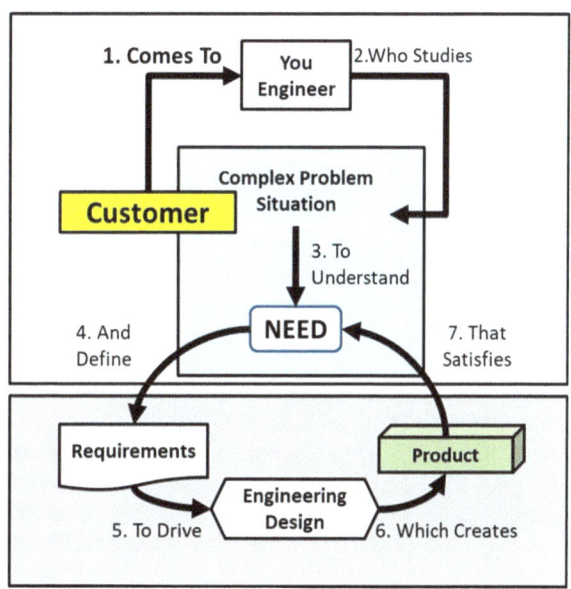

Figure 13: Highlights from Chapter 1

# Chapter 2 - The Systems Perspective is Critical When Designing

*Taking an element out of a system changes it, so when performing engineering design, you must consider elements within the context of the complex problem situation.*

---

Your training as an engineer has prepared you well to conduct the analysis required to design complex components within your field once you know what is needed. It may not have prepared you well to step into a complex problem situation and determine what is required to satisfy the customer's NEEDs. In most of your engineering courses, you utilized the analytical approach to understanding problems. In the analytical approach, first, you disassemble the item you seek to understand. For example, you draw a free body diagram or some other type of model of the pieces. Next, you study and understand each element of the item. Finally, you combine your understanding of the individual elements into an understanding of the whole item. This approach is effective for the detailed design of elements when given a complete set of specifications, but it breaks down when seeking to understand what is required of each element for it to operate within the complex problem situation properly. Taking an element out of the problem situation changes it. Think about the test equipment example on page 5. When designing and only thinking about your test device, it makes sense to perform calibration by simply connecting a calibrated device to your machine. This idea for performing calibration does not make sense when you put your machine back into the problem situation and realize calibration is not performed on site. Putting your design back into the problem situation changes everything because you are dealing with a system instead of individual elements.

When performing engineering design, you do not work with isolated elements. You work with systems, so it is vital that you as an engineer learn to understand and work with systems.

## What is a System?

What do you think of when you hear the word "system"? You may think of items such as a sound system or a car's drive system. Maybe you think of a production system, political system, ecological system, or quality system. Notice that all of the items listed are composed of many parts, but when you think of the item, you do not first think of the parts. You think of the whole because it is a system. This is what a system is. When you think about a system, you typically think about it together, as a whole.

For example, consider a sound system. If you were going to "show" someone your sound system, you would never just take them around and show them all of the elements like the wires, speakers, and power amplifiers. Instead, you would sit them down, put on your favorite tune, and then let them experience the result of all the elements working together. You would let them experience your sound system as a whole. What creates their experience? Is it the speakers? Is it the amplifiers? Is it the wires? No, it is the system. The experience comes from all of the parts working together. The result is more than could be predicted based on the elements alone. This phenomenon is called emergence and is a common characteristic of systems.

With something like a sound system, it is easy to see that you experience the system as a whole and that it makes no sense to modify an element without considering the whole. When performing engineering design, it is easy to forget this fact and focus only on elements removed from the whole. When you are tempted to focus on one element, close your eyes and remember experiencing a great sound system in which all elements work together, and then remember that your element is only one piece of a much larger whole.

Consider another example. When purchasing a new car, you do not go to the lot thinking about nuts, bolts, tires, and belts. You go thinking about the whole, a car. Sure you may look at the details of some of the elements such as the engine horsepower, fuel system efficiency, and cabin volume, but you never look at them in isolation. You look at them in the context of how they contribute to the car (the whole) meeting your needs. The car is made up of many elements, but it makes no sense to think about the elements outside the context of the car. When you take the car for a test drive, you experience the result of all the elements working together to do something none of them can do on their own. Remove or change one element, and the entire experience can change. The car is a system, and systems perform as a whole.

So, what is a system? A general definition of a system is as follows:

> A system is a group of elements interacting and acting as a whole to achieve a common objective that is defined by a wider system.

Notice a few ideas about systems.

- <u>A system is a whole composed of interacting elements.</u> Systems are composed of elements (typically systems themselves), and together the elements create the overall system output (the whole). Elements interact with each other, and each element influences how at least one other element contributes to the overall system output. Recall the sound system and car test drive examples. Every element contributed to the user experiencing the system, and the system output was not just the sum of its elements. The system's parts worked together to create more than any element could create alone.

  Since the system elements all interact and work together, you must make every design decision in the context of the entire system instead of just considering the one individual element on which you may be working. A change to any system element will likely change the user's experience with the system. A positive change to one element may have a negative impact on one or more other elements, so you must consider the entire system when designing an element. You do not want to make your element better at the cost of reducing the overall system performance. For example, consider getting a new digital amplifier for your sound system so that you can take advantage of your digital source devices. You make all of the connections, crank up the volume, press play, and prepare for the ultimate in sound experience. Instead, you hear the poor quality sound, and your amplifier shuts down. Your existing element (4-ohm speakers) interacted with your new element (amplifier designed for 8-ohm speakers) to create an undesired result (amplifier overload).

- <u>Systems are always part of a wider system.</u> Systems are always part of a wider system and can be visualized as forming a hierarchy. Consider your sound system as shown in Figure 14.

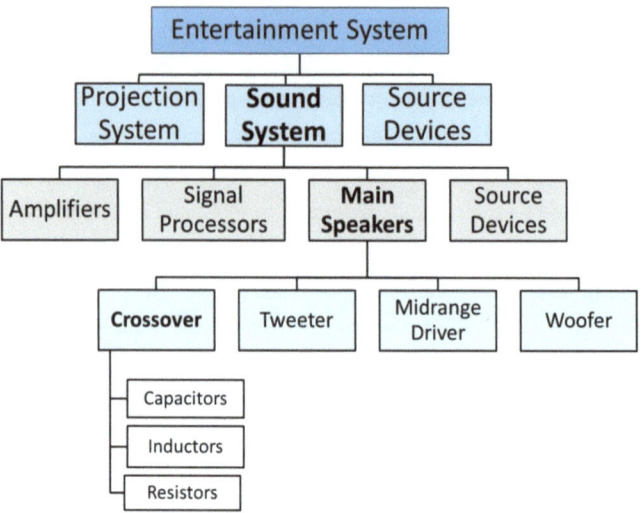

Figure 14: System Hierarchy Example

Your sound system is part of the wider system called an entertainment system. The entertainment system contains elements, and one of the elements is your sound system. Your sound system contains elements, and one element is the main speakers. Each main speaker is a system, and one of its elements is the crossover system. The crossover is made up of capacitors, inductors, and resistors.

For another example of system hierarchy, consider the example of designing a piece of test equipment given on page 5. Figure 15 shows the hierarchy, and your test equipment would be an element of "Gearbox Acceptance Testing" seen in the bottom right corner. The same hierarchy can be represented in diagram form as shown in Figure 16.

The diagram shown in Figure 16 is a form of a context diagram. It shows the item of interest (Gearbox Acceptance Testing System) in the context of the other elements around it. Notice elements in the direct context of the Gearbox Acceptance Testing System such as Gearbox Disassembly, Gearbox Rework, and Gearbox Handling. The interactions with these elements would be given special consideration when designing the Gearbox Acceptance Testing System.

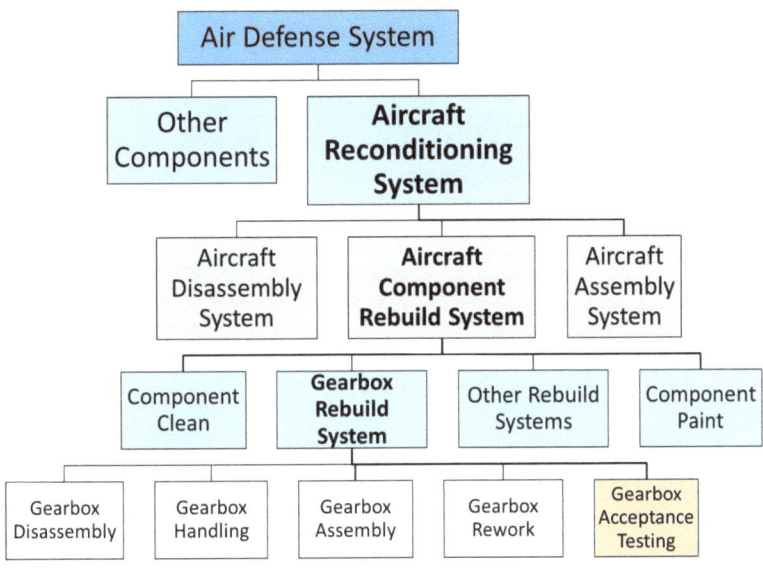

Figure 15: System Hierarchy – Test Stand Example

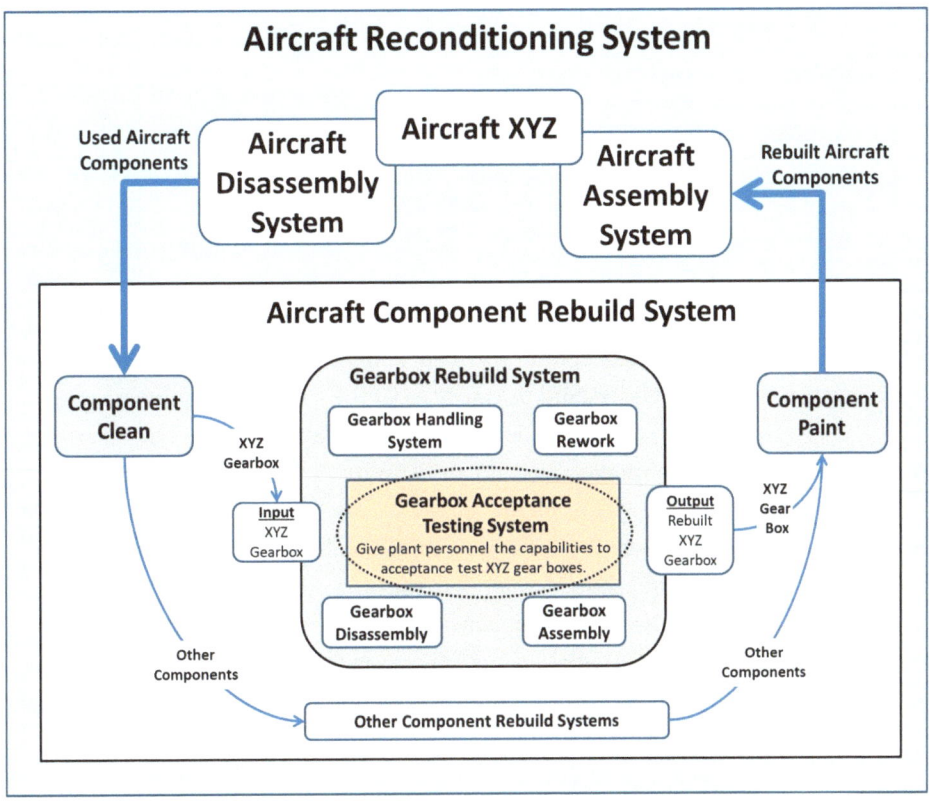

Figure 16: Test Stand Wider System / Context Diagram

The Figure 16 diagram also shows several levels of wider systems. Notice that the Gearbox Acceptance Testing System is part of the wider system Gearbox Rebuild System, and it is part of the wider system Aircraft Component Rebuild System. In this wider system, the Gearbox Rebuild System works with Component Clean, Other Rebuild Systems, and Component Paint to input used aircraft components and output rebuilt aircraft components.

The Aircraft Component Rebuild System is part of the wider system Aircraft Reconditioning System. In this wider system, aircraft components are removed from the aircraft by the Disassembly System, rebuilt in the Aircraft Component Rebuild System, and then placed back on the aircraft by the Assembly System.

When exploring system interactions and needs, it is important to consider several levels of hierarchy as shown in Figure 16. These wider systems define the purpose of your lower level system, and your ultimate goal is to design your system in a way that makes these wider systems better. To truly understand what is needed from (the purpose of) your system, you must understand the wider system.

- <u>A system's purpose is defined by the wider system.</u> The elements in a system work together to achieve a common objective, and this objective (or purpose) always comes from the wider system. Consider your sound system. What do you want out of your sound system? Well, that depends on the purpose of the sound system. The sound system is an element of the entertainment system, and it is expected to fulfill a specific role in the entertainment system. Before you can know the purpose of your sound system, you must understand the purpose of the wider entertainment system. Once you understand the wider entertainment system, you can determine what your sound system must do to support the entertainment system in achieving its purpose. For example, what if the wider system is an outside attraction at Disneyland? The sound system will have very different design needs than a sound system intended for an entertainment system whose wider system is a home theater. The purpose of (what you want out of) the sound system, speakers, and even crossover will be very different depending on which wider system you are accommodating. Understanding the purpose of the wider system is critical when performing engineering design because it defines the purpose of your system.

Consider another example. You are designing a fuel injector for a car. The fuel injector is a system. What do you need out of the fuel injector? What is its purpose? Its purpose depends on the purpose of the car which depends on the wider system to which the car belongs. Obviously what is required from your fuel injector will be very different if the wider system for the car is a NASCAR System, Family Transport System, or a Military Transport System. You cannot truly understand the purpose of your system until you understand the wider system.

Systems exist to achieve some purpose, and the wider system always defines that purpose. A system's purpose is the role it needs to play in the wider system so that the wider system can fulfill its purpose. When performing engineering design, it is vital that you identify and are continually aware of your system's wider system and what it wants out of the system you are designing.

- For design, systems have boundaries. When working with systems, it is important to consider the wider system, but you also must keep in mind elements in the hierarchy that you can and cannot change. Early in a project, you will define what you can and cannot change, and this distinction defines the system boundary for your project. You can make changes to everything within the boundary, and you cannot change anything outside of the boundary. The system boundary is noted in Figure 16 by the dotted line. For this example, the only part of the Aircraft Reconditioning System that can be modified by your design is the Gearbox Acceptance Testing System. You cannot change other system elements in the wider system, but they may interact with your system, so you must consider these other elements when designing.

- System objective (purpose) always rules. Your number one goal must be for the system you are working on to play its role in the wider system properly. You never want to just think about making your element the best it can be. Instead, you want your element to make the system its best. This means that you must always think about the overall wider system and how your decisions impact and are impacted by the other elements in the entire system. You want to optimize system performance instead of your component's performance. You will see this principle in practice when you go through the design example in Chapter 5.

- System elements cannot be completely understood outside of the system. By this point, you should see that taking an element out of the system changes it. The element loses its context and purpose. Systems are not a group of independent elements. Elements within a system interact synergistically, and their performance can only be fully understood when observed within the system. Consider someone looking at your 8-ohm speaker and criticizing you for not utilizing the "better" 4-ohm option. They are looking at your speaker outside of the system. If they looked at the element (speakers) within the system, they would see that your amplifier only accommodates 8-ohm speakers. Their answer obviously would be that you should have selected a "better" amplifier. Again, a look into the system would show that several elements that the system must interface with in the wider system have interface requirements that can only be satisfied by the selected amplifier. Selecting another amplifier would mean that the system would not fulfill its role in the wider system, so your "poor" choice of 8-ohm speakers was truly the best choice for your overall system.

Taking an element out of a system changes it, so when performing engineering design, you must consider elements within the context of the complex problem situation.

These characteristics of a system are summarized in Figure 17.

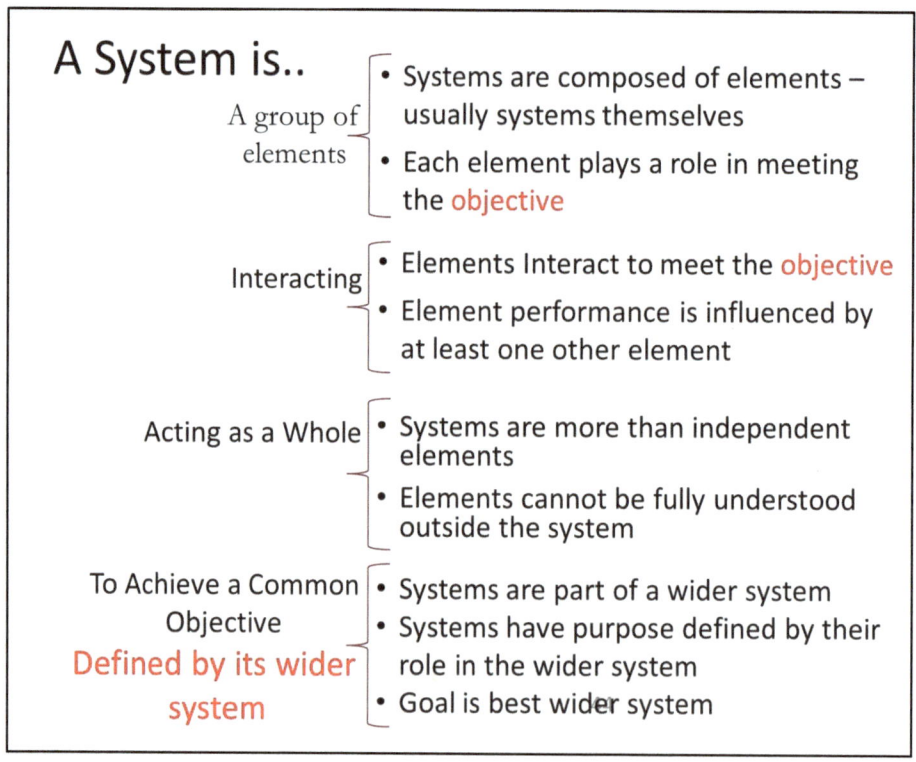

Figure 17: Systems Summary

When you design, you work with systems, and hopefully, you see that the analytical approach by itself breaks down when it removes the component from its context.

To effectively work with the systems within complex problem situations, you need the systems perspective along with your analytical design tools.

**The Systems Perspective**

Based on the unique characteristics of systems, the systems perspective approaches the understanding of problems a little differently from the analytical approach. The systems approach to understanding in the context of design is summarized along with the analytical approach in Figure 18, and it is explained below. I first heard this contrast of the systems perspective and analytical perspective in a speech by Dr. Russell Ackoff.

| <u>Analytical Approach</u> | <u>Systems Approach</u> |
|---|---|
| • Take the item apart | • Identify the wider system which contains the item |
| • Study/Understand the parts | • Understand the wider system |
| • Assemble knowledge of parts into knowledge of what is required for the whole. | • Transfer knowledge of the wider system into requirements for the component to be designed |

**Figure 18: Approach to Understanding**

Instead of taking apart the element to be understood, your first step in the systems approach is to *identify the wider system* that contains that element. Imagine that you are designing a strut to be used in a suspension system for a military transport system. The wider system would be defined as a military transport system.

Instead of taking the element apart, your next step in the systems approach is to *understand the wider system*. For our example, you would need to know information such as the type of vehicle and its use, the terrain the vehicle would drive on, stability required for cargo, how repairs would be made on the vehicle, and how long the vehicle needed to operate between repairs.

Instead of assembling your knowledge of the parts into knowledge of the whole, your final step in the systems approach is to *take your understanding of the wider system and transfer this knowledge to what is required from the element you are trying to understand*. In other words, using your knowledge of the wider system, you define requirements for your element.

Considering the wider system in this manner is the only way to ensure your element will properly fulfill its role in the entire system. For example, assume you determine your military transport system will be carrying cargo that must be kept very stable on all terrains. To satisfy this stability requirement, you determine that you must have an adaptive suspension system. This means your strut must be designed with the ability to vary its stiffness within a specific range and adjust that stiffness within some response time. The only way you could know these specific requirements for your strut is by looking at the wider system.

**Summary**

Engineering design is working with systems, and systems receive their purpose from the wider system. The only way you can fully understand or design a system is by first understanding the wider system and the role your system must play in it. The systems perspective helps you develop this understanding.

# Chapter 3 – Engineering Design

*Engineering design is deliberately determining how to "put something together" so that it will "best" meet the customer's NEED.*

---

You have been asked to design something, but what does that really mean? What is engineering design? I define engineering design as follows:

> Engineering design is deliberately determining how to "put something together" so that it will "best" meet the customer's NEED.

Engineering design is deliberate. Engineering design is not just going to your garage and building something. Engineering design is deliberately planning each detail of the item to most effectively satisfy the customer's NEED. If I point to any feature of your design and ask why it is a certain way, you should be able to tell me. When answering, you should exhibit the discipline of WHAT before HOW. You should be able to tell WHAT the item needed to accomplish, list alternative HOWs you considered, and explain why the selected alternative best satisfies the customer's NEED.

Engineering design is deliberately determining how to put "something" together, and the "something" can be almost anything. We typically think of engineering design in the context of products such as a car, cell phone, or MRI machine, but it is much broader than this. The "something" can be a physical product, a set of procedures, a process, or anything else. If you need something that will contain physical or logical elements that must work together to achieve a common objective, the engineering design process will help you develop an effective solution. I have used it to design a machine to automatically identify defects in wood, an automated factory, a week-long engineering camp, and an engineering program. This list of designed items is diverse, but the WHAT for determining how to put the pieces together to satisfy the customer's NEED is the same for each. Whether you are designing a way to help a non-profit organization better engage with its clients or the next generation of advanced aircraft, the stages you must complete (WHAT) are the same, and these stages can be seen in Figure 19.

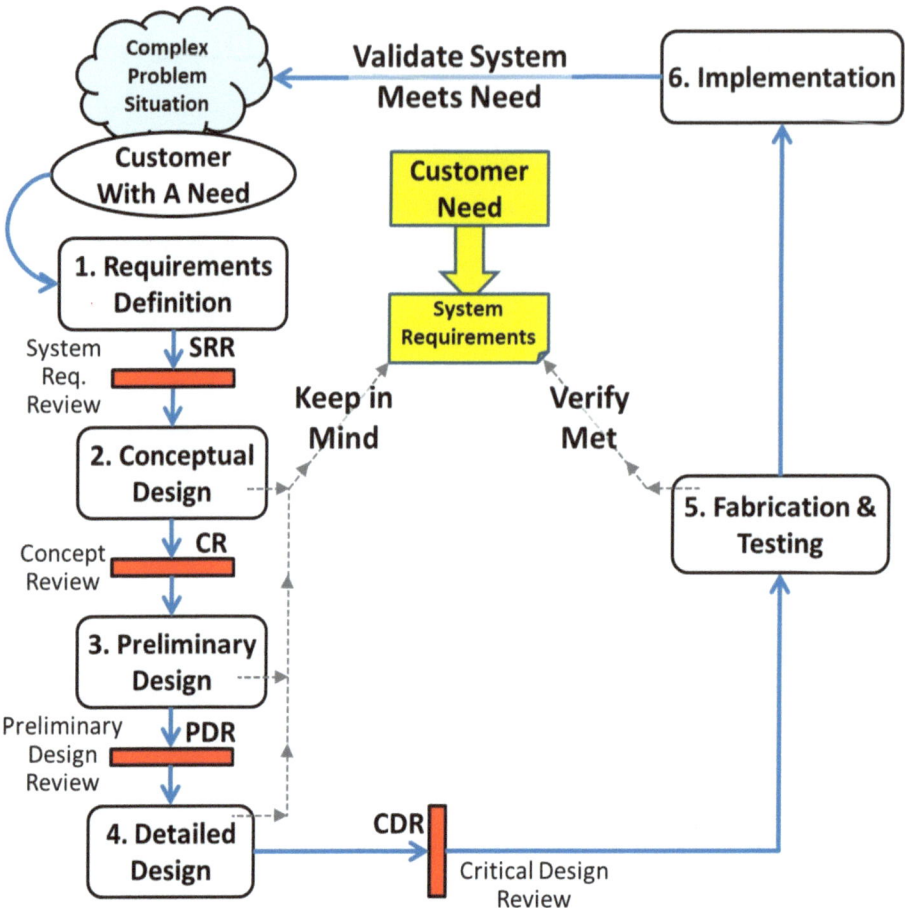

Figure 19: Overview of the Stages in Engineering Design

Figure 19 shows an overview of the typical stages or steps involved in an engineering design project from the perspective of the designer. The process starts with a customer who has a NEED which can be satisfied by a specific product (the physical or logical "something" from above). It ends with the product being implemented in the customer's complex problem situation and satisfying the NEED. Between the NEED and the NEED being satisfied, there are six stages (WHATs) completed for every successful design project. The scope of the problem will drive how much time you spend on each stage and how formally each stage is conducted, but the WHAT for each stage must be accomplished for you to have a successful design. A summary of the stages follows.

- During Stage One (Requirements Definition), you understand the customer's NEED and then express the NEED within a set of requirements.

- During Stages Two, Three, and Four (Conceptual Design, Preliminary Design, Detailed design), you utilize the design thinking cycle explained on page 51 to progressively design a product which completely satisfies all of the requirements. During all three stages, you keep in mind the requirements to ensure that each concept satisfies the customer's NEED. Forgetting about even one requirement can allow you to pursue a design approach that becomes infeasible when the forgotten requirement resurfaces.

- During Stage Five (Fabrication & Testing), the product is fabricated and then verified to ensure it satisfies all of the original requirements. The term fabrication is used loosely here to mean creating the "something" you have designed. For a physical product, this means obtaining each part (purchasing or fabricating) and assembling all of the parts into the "something." When designing a summer camp, fabricating would include activities such as purchasing the materials required for each designed activity, creating the activity handouts as designed, hiring the planned staff, developing a schedule to include all designed elements, and selecting campers based on the designed methodology.

- During Stage Six (Implementation), the product is implemented within the customer's complex problem situation and validated to ensure the original NEED is satisfied.

After each design stage is a formal review - noted in Figure 19 by a red rectangle. Each review has specific objectives, but in general, each review is to prove to the customer and technical experts that you adequately completed the stage and the design is ready to move to the next stage. We all have blind spots, and technical reviews are an opportunity for someone else to review your work and identify details you may have missed. Even if you are designing in an environment that does not have formal technical reviews after each stage, you want to ask some other engineers to review your work from each stage.

Remember, the scope of the problem will drive how much time you spend on each stage and how formally you conduct each phase, but it is necessary to complete each phase for every design problem no matter how large or small.

Figure 19 represents engineering design as a linear series of stages to help you understand the WHAT of designing, but you will not always do the stages in order, and you may revisit a stage several times. For example, when working on Stage Three (preliminary design), you may learn something that makes you go back and work with the customer to refine some requirements developed during Stage One. The work on requirements may cause you to modify your conceptual design (Stage Two), and this will impact your Stage Three work. Figure 19 shows you WHAT goes on when performing engineering design, but you should not view engineering design as just a set of linear steps. Engineering design is an iterative process that involves creativity, analysis, and synthesis. You begin design by looking at the big picture and considering your design as a whole operating within its wider system. Then you focus on various elements or sub-systems of your design and consider the lower level details. You continue going to a greater level of detail until everything about your design is defined. When designing, you often find yourself jumping back and forth between the big picture and the details. The knowledge you gain when working on the details of one part may change your big picture plan of another part. You jump to the impacted level and make some changes. Those changes may impact other items, so you jump to that level and make changes. The back and forth continues until you have your design completely defined.

I view engineering design like an old sailing ship approaching a port for the first time. When still far from shore, the captain can see the land with enough details to identify the port entrance. There is a lot he cannot see at this point, but this does not keep him from beginning the journey to port. Along the way, there may be dangers such as a shallow reef. The captain does not even know to be looking for the reef, but he is aware that he has an incomplete understanding of the situation and that he needs to keep alert for surprises. As the journey towards shore continues, the details become clearer. The captain sees the warning buoys for the reef and now adapts his course accordingly. The closer he gets to his destination, the clearer he sees the details. Every time new details come into focus, the captain adapts his plans until the ship is safe in port. Notice that the captain started off to reach a specific port. The details that emerged during the journey did not change this basic mission. The new details only refined his knowledge of how to best achieve the mission given the details of his current situation.

As a design engineer, you are the captain and a successful design is your port. After requirements definition, you know everything that can be currently known about the problem and about what is required to satisfy the customer's NEED. You do not understand everything, but you know enough to get started in the right direction. As you move forward through the various stages of design, you will stay alert knowing that there is more to learn. Every time new details come into focus, you will adapt your plans until you have a successful design. Like the captain, your basic mission will not change (basic requirements), but as you move through the various stages of design, you will see the problem more clearly, and it is likely that you will need to update some details of the requirements and even modify parts of your design. However, you must be aware. Every change you make will impact other parts of your system. Understanding and managing this change is critical, and it gets harder and more costly 1) the further you are into the design process, and 2) the more people there are on your design team. Like the captain, you follow your course (the engineering design process), and it leads you to port (an effective design solution).

What follows is a description of each stage of the engineering design process from Figure 19. The purpose of this description is not to provide a comprehensive discussion on how you perform design work at each stage. Other authors have given you this detail. Rather, it is to give you a solid understanding of WHAT happens at each stage. HOW you accomplish each stage will change from problem to problem, but WHAT you must accomplish at each stage for an effective design does not change. To be an effective designer, learn the WHAT for each stage, and ensure you accomplish each WHAT for every design you create.

After the discussion on each stage of the design process, the thinking used at every stage of design is presented. You will use this way of thinking at all levels of design.

A case study is used throughout this chapter and in Chapter 4. The case study was developed for use in an undergraduate systems engineering course and utilizes a fictitious company called MixItUp. Before reading the description of each design stage, review the summary of the case study given on the next page.

# Introduction to Case Study

MixItUp has a flexible facility that produces low volume runs of rotational molded plastic parts. Most production runs are low quantities of custom products such as specialty storage containers, pet houses, and pallets. In the manufacturing area, there are three main production cells built around specific capabilities, and each is uniquely configured for the particular product being produced. A conceptual view of the plant is given at the right, and specific cell capabilities are discussed below.

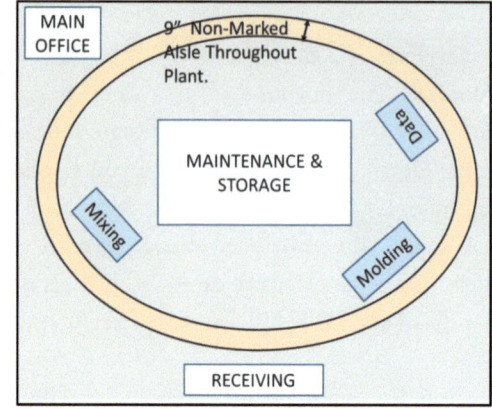

Cell 1 (Molding) contains up to three independent rotational molding stations. Each station has unique capabilities, so not all stations are utilized for each job.

Cell 2 (Mixing) contains an automated mixing station where different types of powder packets are brought and mixed before being delivered to the rotational molding cell. All packets for a specific batch are individually dropped at the cell and automatically moved to one of two queue areas. The cell's main computer interfaces with the office computer to obtain the amount of each type of powder required for each batch. This information can be communicated using Bluetooth.

Cell 3 (Data) is a data collection station. Regulatory agencies associated with MixItUp products require them to keep very strict production records on details such as the types/amount of powder used at mixing and the temperatures of the molding machines. Data is manually collected from the machines and then entered into the company's main computer located at this station.

MixItUp continuously pursues ways to reduce cost, improve quality, and improve worker safety. When studying their operations, two recurring issues surfaced – 1) reliability of the production data industry regulators require, and 2) downtime at the mixing station due to waiting on powder packets. Their internal problem solving efforts led them to conclude that an Automated Guided Vehicle (AGV) can bring improvement in these areas while maintaining (or improving) the plant safety record and the labor required to collect data. They considered using other alternatives to service the cells and concluded that this AGV solution would best meet their needs.

The vision for their solution is that the AGV's "home base" will be at the data collection station where MixItUp will add an automated charging station. The AGV will leave "home base" and travel the plant loop servicing each cell it encounters. After servicing all cells, the AGV will travel back to the data station and transfer the production data collected to the main production computer. After charging, it will return to servicing cells.

MixItUp has come to you to design and build an Automated Guided Vehicle (AGV) to regularly "service" each flexible production cell. Your solution must use the LEGO® NXT (NXT) for any required CPU, and all electronic components must be off the shelf and compatible with the NXT.

## The Typical Stages of Design

The typical stages of engineering design shown in Figure 19 are described below.

Stage One: Requirements Definition

During Stage One, the customer's NEED is fully understood and expressed in requirements. As discussed in Chapter 1, good requirements are critical for an effective design. Requirements guide the entire design process and influence every stage in Figure 19. Requirements allow you to transfer the customer's NEED into a clear and complete picture of what you have to accomplish through your design. You cannot create a design that satisfies the customer's NEED until you understand and can express the NEED. You understand and express the NEED by defining requirements, and these requirements will be passed down to each level of your design. Chapter 4 discusses a specific process you can use to define requirements.

Stage Two: Conceptual Design

Conceptual Design creates the concept for a system that satisfies all of the requirements. During Requirements Definition, you focused on WHAT. During Conceptual Design, you begin to consider HOW. You focus on the big picture WHAT for your system and creatively brainstorm approaches to accomplish the WHAT. You define the elements of your system and how they work together with each other and the wider system to satisfy the customer's NEED. Most design breakthroughs, like the innovative solutions presented in Chapter 1 (page 10), are developed during Conceptual Design. Another term used for Conceptual Design is System Architecting. By determining the big picture elements of your system and defining how they work together, you define your system's architecture or structure.

As you develop design concepts, you keep in mind all of the requirements to ensure that each concept satisfies the customer's NEED. The conceptual design sets the direction for the rest of the design effort, so you carefully consider each idea relative to factors such as risk, cost, ability to satisfy all needs, and other life-cycle concerns designated by the customer such as maintainability, reliability, and manufacturability. For example, if your concept includes a component that requires frequent maintenance, you include features in your conceptual design to facilitate this maintenance.

Eventually, you narrow the ideas down to about two viable alternatives that will completely satisfy all of your requirements. You formally evaluate the different alternatives and then select the best option. You present your results to the customer and work together until you have an acceptable conceptual design. The final design is presented to the customer and a group of technical experts at the Conceptual Design Review.

After Conceptual Design, your design is defined with enough detail to assess that it is feasible and will satisfy all requirements. Figure 20 shows a conceptual design for an automated guided vehicle (AGV) to address the needs presented in the case study. For comparison, representations from the same design will be given after Preliminary Design, Detailed Design, and Fabrication.

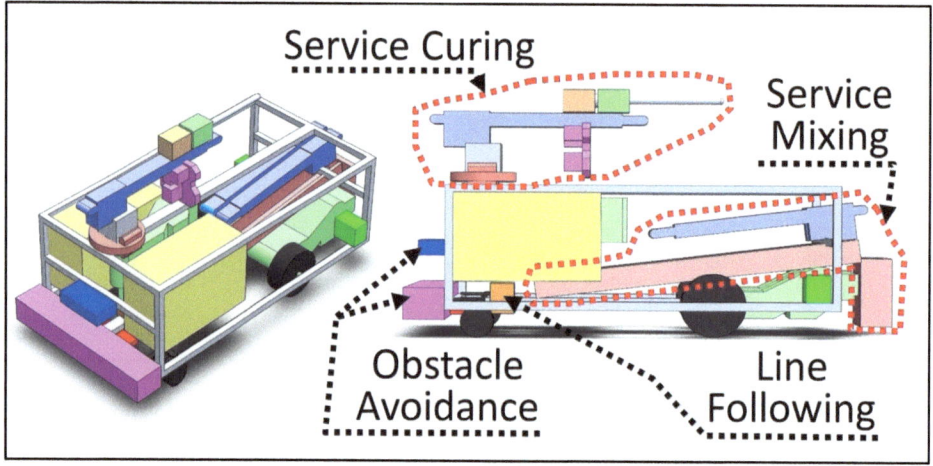

Figure 20: Example Conceptual Design

Notice that most components in Figure 20 are represented as blocks without detail. During conceptual design, you work out conceptually how you will perform each required function, but you do not define every detail of the design. These details will be defined during the later stages of design. The conceptual design focuses on the big picture. You decide what each component will do and how the components will work together and interface with the wider system to satisfy the customer's NEED. Notice some of the decisions made during the conceptual design of the example AGV.

- The AGV will be a traditional four-wheeled vehicle driven by two motors. You may or may not have decided motor details such as the type, and RPM/torque required or the type of wheels.

- You know the general shape and size of the AGV, but specific dimensions for length, width, and height have not been determined.

- The plant floor will be marked with a line, and the vehicle will have a sensor to follow the marked line. You have not decided what type of sensor or line will be used, but you have done enough research to convince the technical experts that it is feasible to use a sensor and marked line to safely navigate the plant.

- The AGV will service the curing cell from the front of the vehicle, and it will service the mixing cell from the back of the vehicle. Notice that this design decision creates a requirement that the AGV must be able to pull into cells forward and backward.

- The design incorporates redundancy for obstacle avoidance (notice two arrows in diagram). You have not selected specific sensors, but you have performed enough research to convince the technical experts that your approach to obstacle avoidance is feasible and will ensure the safety of plant personnel.

Stages Three and Four will fully develop the big picture design developed during Conceptual Design.

<u>Stages Three and Four: Preliminary/Detailed Design–Broad View</u>

You as an engineer have the most experience with Stages Three and Four. Most of the design problems in school start at this point. You have a conceptual design and a clear set of requirements. Your job now is to develop all of the details necessary to turn the concept into a working design that can be built and implemented. You use all of your traditional analytical engineering skills to select components, determine the proper materials, and define details such as the size of beams and the horsepower required for motors. Your final output is a complete design. At a minimum, you will have 1) a complete list of purchased parts, 2) engineering drawings for all manufactured parts, and 3) assembly drawings that define exactly how all of the parts fit together. During the entire design process, you keep the original requirements in mind to ensure your implemented design satisfies the customer's NEED. Remember, forgetting about even one requirement can allow you to pursue a solution approach that becomes infeasible when the forgotten requirement resurfaces (and it will resurface).

Typically, you complete the full design effort in two stages - Preliminary Design and Detailed Design. The difference between the two is the level of detail.

**During preliminary design**, you define approximately 85% of the design details. You select most of the major components of your design and determine how most components work together. You completely define your design from an operational perspective and ensure it satisfies all requirements. Like mentioned with Conceptual Design, you consider and build in design features to address aspects important to the customer such as maintainability, manufacturability, ergonomics, and safety.

Figure 21 shows the example AGV design after Preliminary Design.

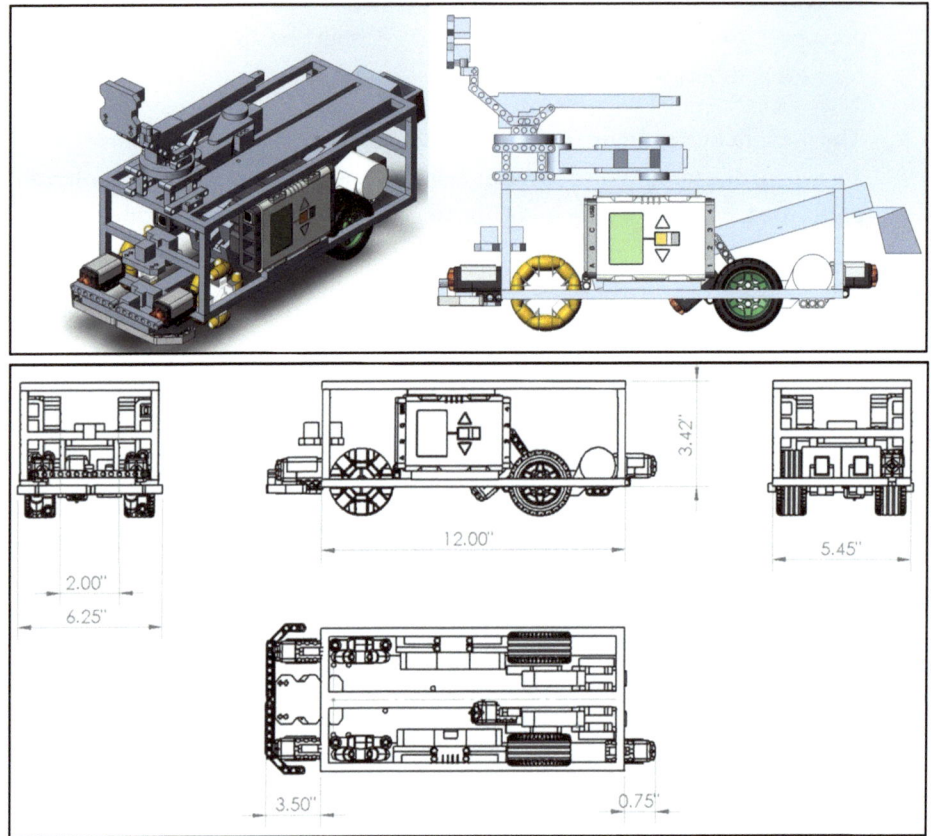

Figure 21: Example Preliminary Design

Notice the following changes in the design from the block diagram shown in Figure 20.

- Motor and wheel alternatives have been evaluated, and specific components have been selected.

- Sensor alternatives have been evaluated, and specific sensors have been selected for most functions.

- Designs for features such as obstacle avoidance have been evaluated, and details of the final design are shown. For example, notice the detail shown for the front bumper with sensors to detect contact with an obstacle. This bumper is the backup for the primary obstacle avoidance method.

- The conceptual design showed rough dimensions for the AGV. The Preliminary design shows specific dimensions for length, width, and height as well as other features of the design.

Preliminary Design ends with the Preliminary Design Review (PDR). At PDR, you present the details of your design to the customer and technical experts. The customer is there to ensure your design satisfies all of the requirements. The technical experts are there to review the technical aspects of your design and ensure it will operate as desired when built. View these experts as someone with more engineering experience than you looking over your shoulder to ensure you have not missed anything in your design and to provide insight based on their technical experience. This advisory role is sometimes called technical conscience. They will also evaluate your design against performance parameters important to the customer such as cost, safety, risk, usability, ergonomics, reliability, and manufacturability.

Once the design moves to Detailed Design, a large amount of engineering labor is expended, and design changes become very expensive. To avoid costly mistakes, it is imperative that you thoroughly review your design and identify any needed changes before you move to detailed design.

**During Detailed Design**, you refine your preliminary design and completely define your design from an engineering perspective. Every detail of every aspect of the design is defined in a way that allows it to be fabricated. You use the techniques learned in your engineering fundamentals courses such as strength analysis, vibrations analysis, and heat transfer to ensure every detail of every component is properly designed for a safe and effective design. Purchased parts are specified. Fabricated parts are defined with engineering drawings. Assembly drawings define exactly how all of the parts fit together. The first time you perform detailed design, you will be amazed at 1) how many details must be defined, 2) the time it takes to define all of the details, 3) the time it takes to document your design in detailed drawings, and 4) the effort involved in locating sources for and specifying every purchased part.

For an example of the work done during detailed design, consider the example AGV design.

- During Preliminary Design, you designed the front bumper system and determined exactly where it would be located. You did not determine details for the bumper such as the type of material to use, how large the material should be to provide the required strength, and how the bumper will attach to the AGV frame. During Detailed Design, you determine every detail of how it will attach. Will the bumper be welded, bolted, or attached using some other method? If bolted, you will perform engineering analysis to define details such as the type of bolt to use, the size and thread of the bolt, the type of head on the bolt, and the specific bolt pattern to provide the required stability of the bumper.

- During Preliminary Design, you decided that the plant floor will be marked with black tape and a multi-sensor device will be used to determine the position of the AGV relative to the line. During Detailed Design, you must answer questions such as how thick will the line be, what kind of tape will be used, what kind of control algorithm will you use to keep the AGV on the line, and how will you know when the line is lost?

- During Preliminary Design, you designed the chute to deliver packets at mixing and defined its approximate angle of tilt. During Detailed Design, you define exactly how the chute will attach as well as the exact angle required for proper delivery of packets.

Detailed Design ends with the Critical Design Review (CDR). At CDR, the technical experts review all aspects of your design to ensure it is technically sound and will satisfy all customer requirements. They also review your design and your documentation (drawings) to ensure everything is ready for fabrication.

Figure 22 and Figure 23 contain examples of two parts from the AGV example after Detailed Design.

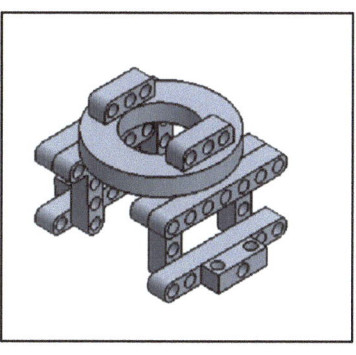

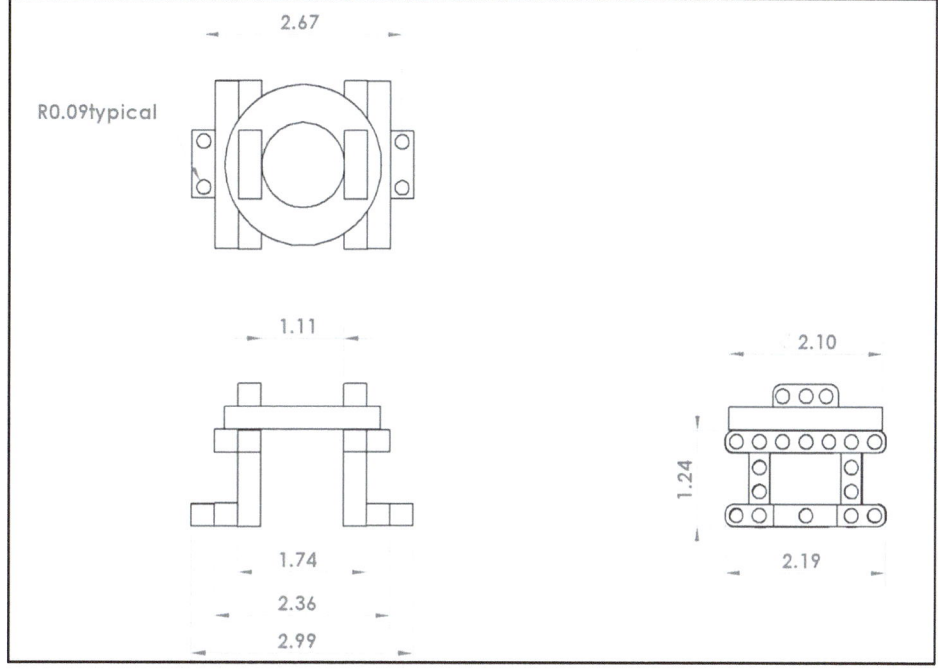

Figure 22: Example of Detailed Design #1

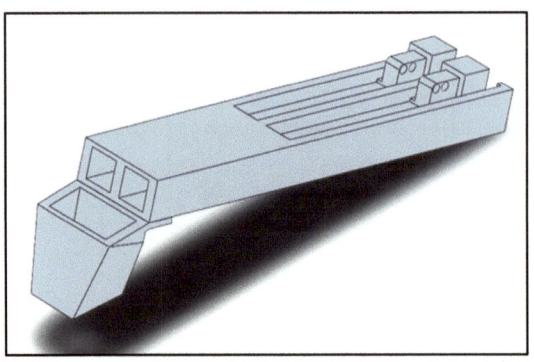

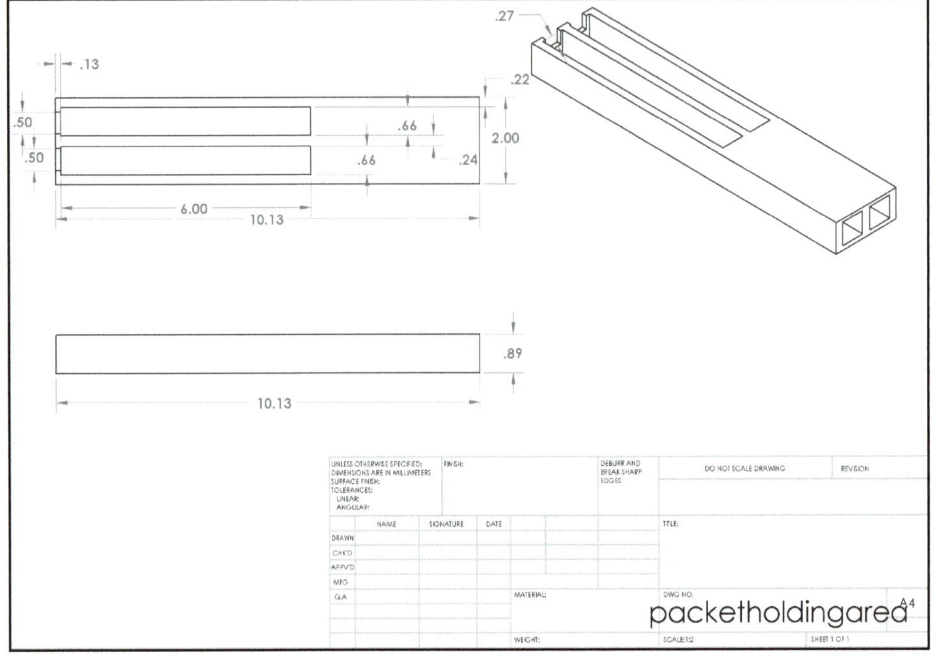

Figure 23: Example Detailed Design #2

Stages Three and Four: Preliminary/Detailed Design–Detailed View

Before we discuss Stage Five (Fabrication), we need to take a more detailed look at what really happens when performing preliminary and detailed design.

Due to the complexity of most design efforts, the overall system is typically split into several sub-systems at some point during the Conceptual Design stage. For example, the case study AGV conceptual design shown in Figure 20 is divided into the Curing Sub-System, the Mixing Sub-System, and the Drive Sub-System as shown in Figure 24.

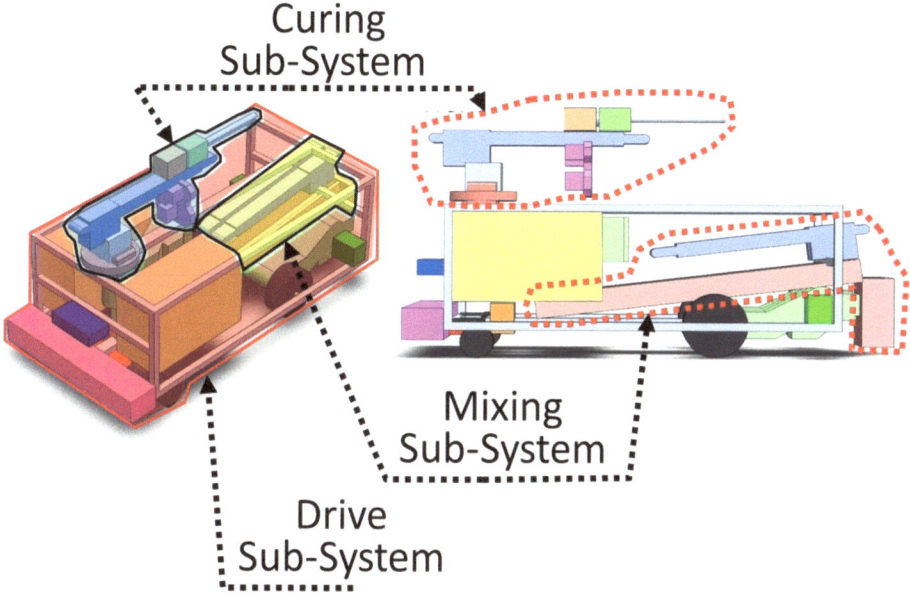

Figure 24: Sub-Systems for Case Study Example Design

Each sub-system is responsible for accomplishing specific functions, and a team is given responsibility for the design of the sub-system. In our example, the Drive Sub-System is responsible for navigating the plant aisle, identifying cells, maneuvering in and out of cells, and avoiding obstacles. The Drive Team completes Stages Three (Preliminary Design) and Four (Detailed Design) as described above to create a design for the Drive Sub-System. At the same time, the Mixing Team and the Curing Team work on their sub-systems and complete Stages Three and Four. To ensure the details of each sub-system are properly integrated to satisfy the customer's NEED, a team is given responsibility for the overall system. The Integration Team facilitates communications to ensure all of the sub-systems properly work together and satisfy the customer's NEED. A detailed view of what happens when performing preliminary and detailed design is shown in Figure 25.

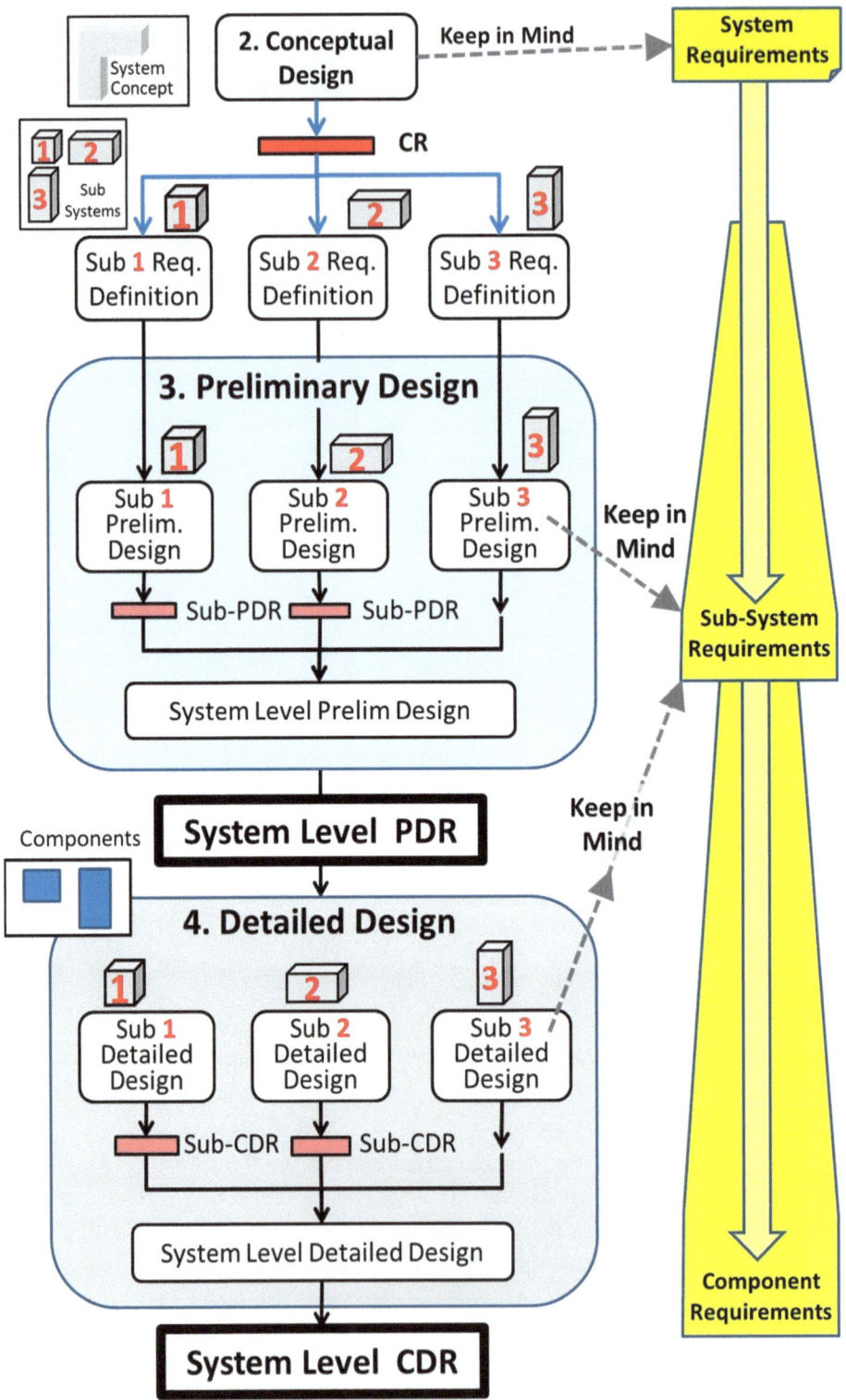

Figure 25: Detailed View of Preliminary and Detailed Design Stages

Figure 19 and the accompanying discussion gives you a good idea of what you accomplish during each stage of design. Figure 25 provides details on the tasks between Stage Two (Conceptual Design) and Stage Five (Fabrication and Testing) to show how you typically perform the stages. Notice the following in Figure 25.

### Definition of Sub-Systems

During Conceptual design, you focus on the overall system concept, but at some point, you divide the system into sub-systems. This division is helpful and necessary, but do not miss the fact that dividing the overall system adds complexity to your design effort. Consider the three sub-systems for the AGV shown in Figure 24. Even without knowledge of the systems perspective, it is obvious that the three sub-systems are interconnected and impact each other. The Drive sub-system must physically accommodate the other two sub-systems, and both sub-systems add weight to the Drive sub-system. This weight will likely impact the speed and maneuverability of the Drive sub-system. The alignment of the Curing sub-system and the Mixing sub-system in reference to a particular cell is determined by the Drive sub-system's alignment of the AGV within the cell. Everything discussed concerning systems in Chapter 2 now applies between your sub-systems. Just like you carefully consider how your system interfaces with entities in its wider system, now each sub-system design team also considers the interfaces each sub-system has with the other sub-systems.

Sub-system interfaces are largely driven by how the system is divided into sub-systems, and this task is part of system architecting. The details of systems architecting are out of the scope of this book, but please note that your definition of sub-systems can increase or decrease the complexity of the interfaces between sub-systems, and this definition can impact many operational aspects of your completed system. Carefully consider these internal interfaces when defining your sub-systems.

### Development of Sub-System Requirements.

Between Stage Two (Conceptual Design) and Stage Three (Preliminary Design) notice that there is a block to define requirements for each sub-system. During Stage One shown in Figure 19, you understood the customer's NEED and expressed this NEED in a set of requirements for the system. Now, this NEED must be expressed for each sub-system in the form of sub-system requirements. Most of your sub-system requirements will come directly from the system requirements, so you define all of these requirements now. Other sub-system requirements will come from interfaces with other sub-systems. For example, the Drive sub-system will have a requirement that it must carry a certain amount of weight. This weight will be defined in part by the weight of the two sub-systems it must carry. These interface type sub-system requirements cannot all be defined at this point. Instead, you will define them as the sub-system designs progress through Preliminary Design. This definition of sub-system requirements is shown on the right in Figure 25. Notice that the definition of sub-system requirements starts before Preliminary Design and continues to the end of Preliminary Design. The Integration Team ensures all sub-system design teams are coordinating their designs and mutually defining interface sub-system requirements.

**Sub-System Preliminary and Detailed Design.**

Notice in Figure 19 that the blocks for Stage Three (Preliminary Design) and Stage Four (Detailed Design) both show individual blocks for the design of each sub-system based on each sub-system's requirements. Each sub-system design team is careful to keep all of their sub-system requirements in mind as their design progresses. As always, they consider interactions between their sub-system and the wider system elements, but now they also consider interactions with the other sub-systems. These interactions are easy to forget, but the Integration Team helps each sub-system design team consciously consider the sub-systems they interface with and coordinate with the other design teams. For large sub-systems (sub-system 1 and 2 in Figure 25) a sub-system Preliminary/Detailed Design Review is conducted before the system level review.

At the end of Preliminary/Detailed Design, all of the sub-system designs are brought together to form the system level preliminary/detailed design. This design is presented in the Preliminary/Detailed Design Review.

**Linking the Design to the Customer's NEED**

Notice the left side of Figure 25. The focus of your design effort begins with the overall system and then shifts to each sub-system. As the design progresses through Preliminary Design and into Detailed Design, your focus moves to the individual components that comprise each sub-system. The stages of design are structured to help you integrate the customer's NEED into each level of your design.

Consider the requirements on the right side of Figure 25. The customer's NEED is expressed in the System Requirements. These system-level requirements drive your Conceptual Design to ensure the system concept developed satisfies the customer's NEED. As focus moves from the overall system to the sub-systems, the customer's NEED represented in the System Requirements is passed to each sub-system design team through the sub-system requirements. As you perform Preliminary Design and Detailed Design, you keep all of the sub-system requirements in mind to ensure that each aspect of your design satisfies the customer's NEED.

During Detailed Design, you define the component requirements necessary to ensure each component performs as needed to accomplish the sub-system requirements.

In summary, follow the right side of Figure 25 from the top down. The customer's NEED is represented in the System Requirements. The System Requirements are passed to the Sub-System Requirements. The Sub-System Requirements drive the design of each sub-system and component. The detailed design of sub-systems creates Component Requirements. The Component Requirements specify exactly what is necessary for each part to accomplish the Sub-System Requirements which in turn will meet the System Requirements. Meeting the System Requirements should satisfy the customer's NEED.

Stage Five: Fabrication and Testing

It is now time to build and test your system. You spent a good bit of time up front developing requirements that your system must satisfy to meet the customer's NEED, so you check each part of your design to ensure it satisfies all requirements before assembly. As your design is created, you verify each requirement as described below.

- First, you gather all of the components (purchased or manufactured) and verify that they satisfy the Component Requirements. For example, Figure 26 shows the fabricated Mixing Sub-System component from Figure 23. You check this component against all critical dimensions in the design drawing to ensure it will perform as needed in the overall design.

Figure 26: Component of Mixing Sub-System

- Once you know your components meet your requirements, you assemble them into their respective sub-systems and then verify that each sub-system satisfies all Sub-System Requirements. For example, Figure 27, shows the completed Mixing Sub-System. Before being assembled into the full system, you thoroughly test this sub-system to ensure it satisfies all Sub-System Requirements.

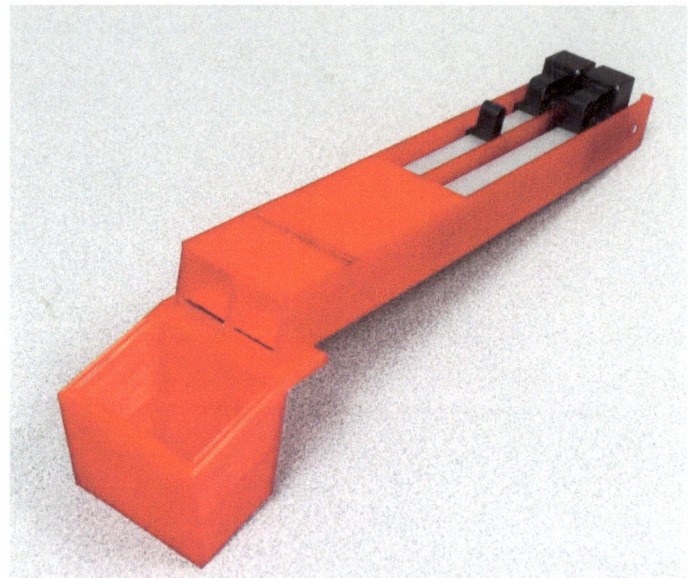

Figure 27: Mixing Sub-System

- Once you know all sub-systems perform as needed (you have verified all Sub-System Requirements), you assemble the sub-systems into the full system. The complete fabricated system for the design shown in Figure 20 (Conceptual Design) and Figure 21 (Preliminary Design) is shown in Figure 28.

Figure 28: Example Fabricated Complete System

- After the entire design is fabricated and assembled, it is thoroughly tested to verify that it satisfies each of the System Requirements. Modifications and improvements are made as necessary. When testing is complete, you know your system satisfies all of the requirements you defined. It has been built as designed, but is it the correct system that will truly satisfy the customer's NEED? This question is answered in the final step.

Stage Six: Implementation

The final step in the design process is implementing the system in the complex problem situation and validating that it truly does satisfy the customer's NEED. Now is the moment of truth. You install the system in the complex problem situation and perform more testing to ensure that it does, in fact, satisfy the customer's original NEED. You verified that the system was built correctly from the previous step. Now you are validating that you built the right system. When all criteria are met, the system is accepted, and you receive final payment from the customer.

Figure 29 provides a final view of the stages of design. Use this figure as a reminder during your design project. The middle of the figure reminds you how the process takes the customer's NEED and embeds it into your design through System Requirements, Sub-System Requirements, and Component Requirements. The left side of the figure reminds you of the key stages you should perform to create your design and highlights how you must keep the System Requirements and Sub-System Requirements in mind at every stage of your design. The right side of the figure reminds you to check each component, sub-system, and system against the proper requirements before moving up to the next level of fabrication. Following these steps ultimately helps you satisfy your customer's NEED.

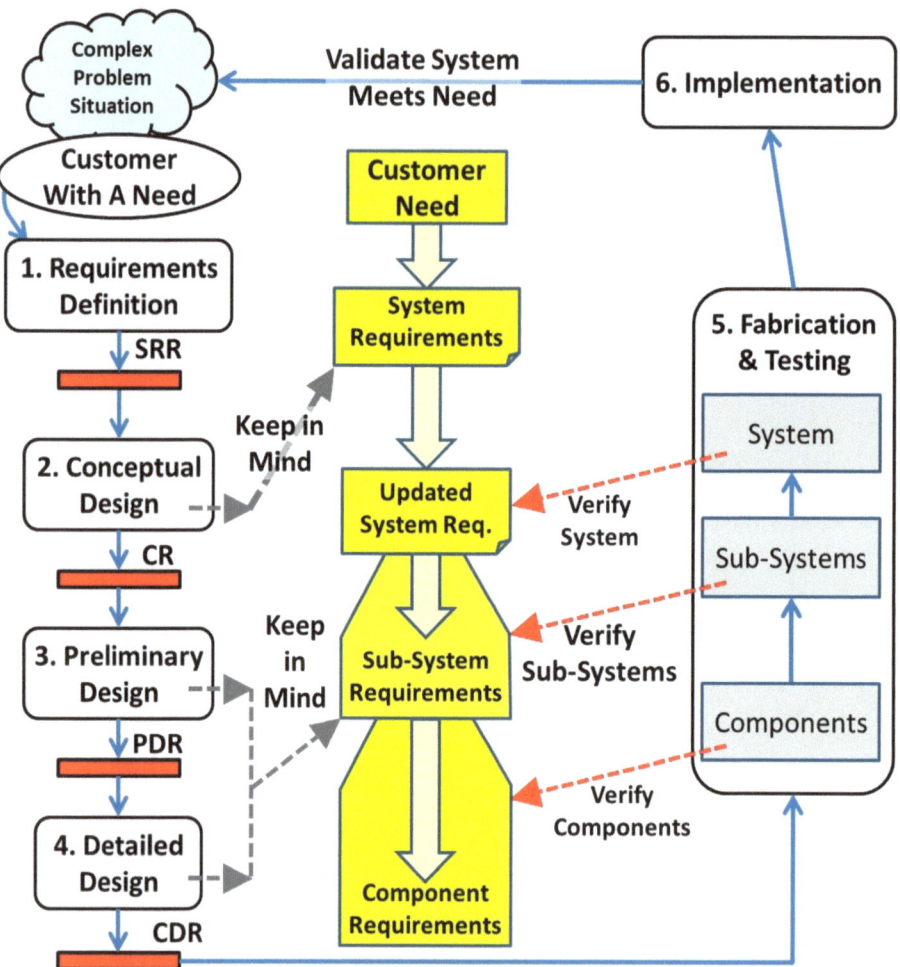

Figure 29: Stages of Engineering Design with Verification Steps

<u>After Implementation</u>

Figure 29 shows the typical steps for an engineering design project from the perspective of the designer, but you need to remember two additional steps – Sustainment and Disposal. The implemented system will operate for some amount of time in the complex problem situation, and it must be sustained during this time with routine maintenance and repairs. At some point, the system will no longer be needed and will be retired. Decisions you make during design can have a significant impact on sustainment and disposal costs. Effective designers consider these costs during the initial design of the product.

**Design Thinking**

Figure 29 shows the various stages of design, but what do you do when performing each of the stages of design? How do you design? What do you do when performing Conceptual Design, or Preliminary Design, or Detailed Design? Many tools facilitate design, but the essence of engineering design is applying the thinking illustrated in Figure 30 over and over at various levels of detail. Recall that design is deliberately determining all of the details for the item being designed. Figure 30 represents the thinking used to determine each detail.

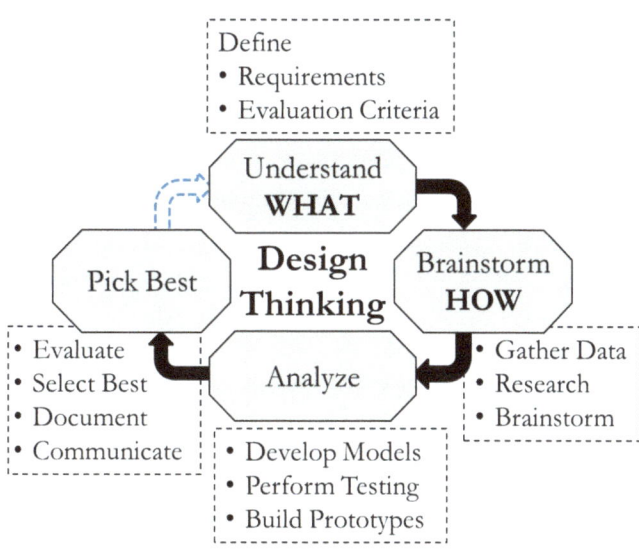

Figure 30: Design Thinking

I call the thinking represented in Figure 30 "Design Thinking," and it is used at every level of design. You use it during Conceptual Design to develop your broad system concept, and you use it during Detailed Design to determine the best way to attach one component to another. Design Thinking is composed of the following four steps. Step 1: First, you define WHAT you are trying to accomplish (Requirements) and how you will judge which idea is better than another (Evaluation Criteria). If you are doing high-level conceptual design, the WHAT will be the NEED you defined during Requirements Definition. If you are doing a low-level design of a sub-system, the WHAT may be a problem you are trying to solve like "stabilize the steering bar."

Step 2: Once you know WHAT you are trying to accomplish, you will usually have to do some research and learn about the topic at hand. You also may have to gather data from other parts of your design to understand the situation and know exactly what is needed out of the element in which you are working. After learning about the topic, you brainstorm alternative ways (HOWs) to accomplish the WHAT. Sometimes you will start brainstorming and realize that you still do not know enough to generate viable ideas effectively. In this case, you stop, perform research or gather facts, and then come back to brainstorming.

Step 3: After brainstorming, you analyze each idea and select the one that best satisfies your evaluation criteria. Selecting the best idea often requires much work. Remember, you are an engineer, and your decision must be based on facts about each idea. You do not pick an idea based on your opinions. You pick the alternative that the facts show will best satisfy your requirements. Gathering the necessary facts means that you will likely have to perform some more research on the various alternatives. You may have to perform some calculations. You may have to build some models and perform testing. You may have to build and test some prototypes of your design (or parts of your design). Whatever you do to analyze the alternatives, you always keep your eyes on the requirements and pick the alternative that best satisfies them.

Step 4: After doing whatever is necessary to gather the facts required for making an informed decision, you select the best alternative. The design decision is documented and communicated to the rest of the design team, and now you are ready to do the process all over again for the next WHAT. This process continues until every detail of your design is defined to satisfy the requirements.

**Summary**

Chapter 1 opened with you being asked to design something for someone. Hopefully, you now have a much broader view of your role as an engineer and realize that you are a "Needs Satisfier" instead of just a designer. This chapter opened with the question "What is engineering design?" You should now understand that design is a process and a way of thinking. The process is a series of stages which help you embed the customer's NEED into every aspect of your design. The thinking (Design

Thinking) is an extension of the Discipline of WHAT before HOW explained in Chapter 1, and it helps you deliberately determine the best alternative for every decision you make when designing.

Engineering Design is a very detailed activity, and it is so easy to get consumed with the technical details of your particular sub-system; especially during Detailed Design. When working on the details of your design, periodically stop and look up from the details. Look around and remember that you are in a system. Remember all of the points made about a system on page 21 and that your sub-system is only a part of the full system. Remember Design Thinking and review your Sub-System Requirements often. Remember to consider the other sub-system design teams and to make sure your details fit in with all of the other parts. In all of this, remember that you must consider the wider system as well as the other sub-systems that make up your design. Make sure you are communicating with all other elements and that you take the interactions with them into account with your design.

Chapter 3 will provide details on how to understand the customer's NEED and develop requirements. You will learn steps to take and tools to use in defining requirements. Like anything, it will be easy to get lost in the details of how to define requirements. To help you not lose the key ideas in the details, the systems perspective for design has been condensed into four cardinal rules for an effective design. These rules can be seen in Figure 31 and should often be reviewed during your design projects.

## 4 Cardinal Rules for an Effective Design

1. Understand the wider system that contains what you are designing (your system)

   - Identify the wider system
   - Understand the wider system
   - Define exactly WHAT the wider system wants out of your system
   - Transfer your understanding into requirements for your system

2. Understand the other elements in the system that interact with your system

   - Identify the elements
   - Understand the other elements
     - What do they want from your system?
     - How does your system influence/interact with them?
   - Transfer your understanding into requirements for your system

3. Design to create the best wider system and not just to make your system best.

4. From time to time, lift your head from the details and go back to the other steps.

Figure 31: Summary of the Systems Perspective for Design

# Chapter 4 - Five Steps You Can Use to Define Requirements

*Defining requirements is understanding and defining everything necessary to satisfy the customer's NEED.*

*Requirements are the specific wants and needs that you and the customer agree will be addressed by your system to satisfy their NEED.*

---

Before I discuss steps you can use to define requirements, I want to review something from Chapter One. Recall that you will discover *wants* and *needs* when defining requirements. A need is something essential to meeting the NEED. A want is something that may be helpful, but technically the NEED can be satisfied without it. On paper, it is easy to make a clear distinction between wants and needs, but it is not so easy when working in the complex problem situation. Often, it is hard to distinguish between a need and a strong want, and sometimes a customer will not be satisfied unless they receive a want. Since the customer will not be satisfied without the want, does this make the want a need? We can discuss this question for a very long time without reaching a definitive answer, so I take the following approach when defining requirements.

- When exploring what is necessary to satisfy the customer's NEED, you are **always** identifying wants and needs, even though you may not always be sure which is which.

- Since you are always looking for wants and needs, I use terms like "needed," "needs," and "necessary" loosely to include wants and needs. When I use these words in the context of defining requirements, I am referring to wants and needs. For example, when I use phrases like below, I am always referring to wants and needs.

    "Define everything necessary to satisfy the NEED."

    "Determine all of the needs."

    "Determine what is needed."

Once all of the needs (remember, I mean wants and needs) are understood, they are reviewed to determine the specific needs (wants and needs) to be addressed by your design. These needs are refined to form the requirements.

In this book, requirements are the specific wants and needs that you and the customer agree will be addressed by your system to satisfy their NEED.

There are many ways to define requirements, and as you mature as a designer, you will develop your unique method. In the meantime, you can use the five steps shown in Figure 32 and presented in this chapter.

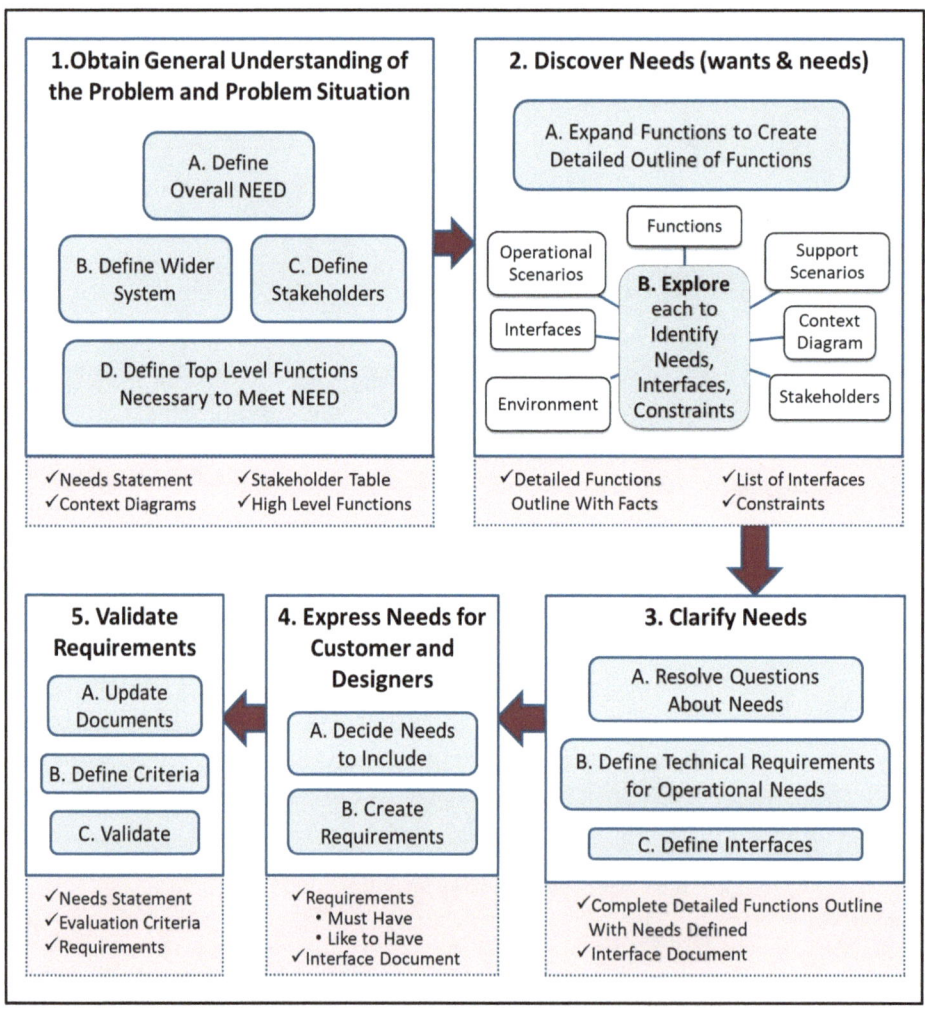

Figure 32: Steps to Develop Requirements

To define requirements, you initially develop a general understanding of the customer's problem and the wider system (Step 1). After you generally know WHAT the customer wants you to accomplish, you explore the situation in detail to discover everything required to satisfy the customer's NEED (Step 2). To clarify your understanding of the needs, you resolve any unanswered questions and, where necessary, turn customer operational needs into technical requirements (Step 3). You work with the customer to define the specific needs to be addressed by your system and then format them to form the system requirements (Step 4). Finally, you define evaluation criteria to use when making design tradeoffs and formally present the requirements to the customer for validation (Step 5).

The five steps in Figure 32 build on the ideas presented in "Getting to Requirements: the W5H Challenge" by Gaasbeek and Martin. Review this article to gain more insights into the process of defining requirements.
(http://onlinelibrary.wiley.com/doi/10.1002/j.2334-5837.2001.tb02413.x/pdf).

The steps also utilize several tools presented by Dr. Stuart Burge. Reviewing these tools on Dr. Burge's website when they are used in the steps will be helpful.
(https://www.burgehugheswalsh.co.uk/Systems-Engineering/Tools.aspx).

**General Principles to Follow When Defining Requirements**

At first, defining requirements can seem bit overwhelming, and it is easy to lose your way in the process. To stay on track, keep the following general principles in mind.

Never lose sight of the WHAT behind defining requirements. All of the work you do in the five steps is to accomplish one thing; understand everything necessary to satisfy the customer's NEED.

Like the process of designing, defining requirements is represented as a series of steps (Figure 32), but the process of defining requirements is not linear. The steps define WHAT you do to define requirements and a general order to do it in. Review the ship approaching shore analogy on page 32. You will begin defining requirements with very little knowledge of the system (very far from shore). The initial artifacts you create, such as the Needs Statement or Detailed Functions with Facts, will represent your current limited understanding. As you progress through the process of defining requirements (move closer to shore), your knowledge of the system will increase and you will update all of your previous artifacts as it does. When you reach the end of defining requirements (you are not on shore yet, only done with Stage One of design), all of your artifacts will represent your current knowledge of the system. As you progress through the other stages of design, your knowledge of the system will continue to increase, and you should update your artifacts. As this happens, remember my previous warning. Every change you make may impact other parts of your system. Understanding and managing this change is critical, and it gets harder and more costly 1) the further you are into the design process, and 2) the more people there are on your design team.

Do not get lost in the creation of artifacts, especially visuals. When defining requirements, you will create many visuals, but you must remember your WHAT. Learning about the system and what is needed to satisfy the customer's NEED is your WHAT, and the visuals are just a tool to help you understand the system and the customer's NEED. Specifically, the visuals will help you in the following ways.

- Initially, creating a visual will help you personally, because it forces you to explore and understand the situation well enough to create it. Practically speaking, some visuals will have no other purpose than facilitating your understanding through its creation.
- Throughout the design project, visuals will help you maintain the systems perspective. Periodically reviewing your initial visuals is a quick way to be reminded of the wider system and the key interactions within the system.
- Once created, some visuals help you to communicate essential details that you learned to others who have not studied the system as you have. Remember, the visual is a communications tool, so adapt it to communicate what is needed for your situation. Also be aware that some visuals will only be helpful to you and they will make no sense to anyone else.

In this chapter, the steps to develop requirements are presented as something you individually complete, but in reality, you are working with the customer throughout the entire process. Recall from Chapter One that the customer does not fully understand the system or their NEED. As you understand the system and the customer's true NEED, you should bring them along with you, so they understand the system and their NEED more clearly as well. The customer has most of the information you need to define requirements, but they likely do not know it. Your job is to provide the right experience and ask the correct questions to draw this knowledge out. The five steps shown in Figure 32 are designed to help you do this with the customer.

Understanding the customer's need is all about information, and you will acquire information concerning needs from many sources. Some of the common ways you will learn are given in Figure 33. Stop now and review these sources of information. None of the sources will give you complete information, and you will find yourself going back and forth between many of the sources. For example, what you hear from a stakeholder may cause you to research a particular national code. What you learn about the code may cause you to talk more with the stakeholder.

> **Knowledge concerning needs comes from:**
>
> - Using your own background and experiences to think about the system and potential needs
> - Spending time observing the various elements within the problem situation
> - Talking with stakeholders and others within the problem situation
> - Reviewing documents such as company memos, equipment manuals, local and national code publications
> - Facilitating meetings designed to explore needs

Figure 33: Key Sources of Information on Needs

Okay, I think I have prepared you enough. The remainder of this chapter will provide a detailed explanation of each of the five steps to defining requirements and will instruct you to begin defining requirements for your problem. I recommend you read the chapter all the way through and then go back through it again and follow the steps for your design problem. Good luck!

**Step 1: Obtain General Understanding of the Problem & Problem Situation**

During Step 1, you are trying to develop a broad view of what the customer is really asking for and of the general context of the problem situation. The customer came to you because they have a problem (NEED). What is it? Typically the customer will have a general idea or vision of a solution or solution approach they want you to use. After you understand their NEED and problem situation, you will evaluate if what they are asking for will truly satisfy their NEED. If it will, then you will move on to Step 2. If what they are asking for will not meet their NEED, you will discuss your conclusions with the customer and agree on a constructive path forward.

The four actions you will take during Step 1 can be seen in Figure 34, and a description of each action follows.

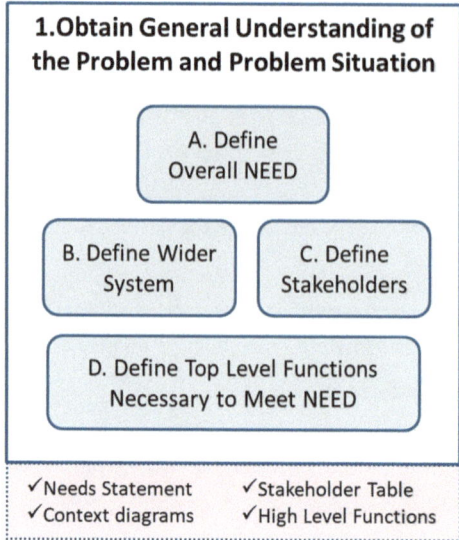

Figure 34: Tasks for Step 1

Action 1-A: Define Overall NEED

Your first task is to gain an overall understanding of the problem situation and the customer's NEED. You should meet with the customer and learn about the situation from their perspective, but you also want to spend time in the problem situation where your system will be implemented. Familiarize yourself with the situation by visiting the area where your system will operate, looking at the other elements in your wider system, reviewing relevant documents, and talking with those working in the problem situation.

For your design problem, continue to explore the problem situation until you can answer the questions below.

- Why does the customer want the new system?
- How does the customer think the new system will help them?
- What is the wider system that the new system will operate in?
- What role will the new system play in the wider system?
- What are the current deficiencies (gaps) the customer is trying to fill with the system? If there was a previous system, what is the need for a change?
- What is important to the customer?
- What is the customer's vision of how the new system will operate?
- Where will the new system operate and under what conditions?

- What restrictions or limitations are involved in the project? Are there any standards (company, local, federal) that must be followed?

- Who will operate the new system? Who will maintain/fix the new system? Will special training be required?

Typically Action 1-A and 1-B are completed somewhat together. As you understand the problem situation and explore answers to the questions above, developing the Context Diagrams in Action 1-B helps clarify your understand and will generate additional questions to explore. Working on the context diagrams will give insights which will help answer the questions. Continue exploring the system until you create the context diagrams and answer the questions.

Once you can answer the questions above, you will summarize your understanding of the Customer's NEED by developing a draft Needs Statement for the project as shown in Figure 35 for the case study.

---

**MixItUp**
Develop a device that can safely operate unassisted to service the MixItUp production cells.

---

**Figure 35: Example Needs Statements**

The Needs Statement is a short description that captures the essence of your system. It should be a technology-free statement that focuses on WHAT and not HOW. The Needs Statement is an "elevator speech" for your system and will be used throughout the project to help keep everyone involved focused on WHAT you are trying to accomplish. It is very easy to get lost in the details of your design and lose sight of the true NEED you are satisfying. The Needs Statement helps this not to happen. Even a new member of the project team should be able to look at the Needs Statement and understand the big picture of what the team is trying to accomplish. For example, review the MixItUp Needs Statement shown in Figure 35 for the case study. Anyone reading the Needs Statement would immediately know that your device needs to be safe, operate unassisted, and service the MixItUp production cells.

Stop now and create a short description that captures the essence of your design project. You will refine this draft Needs Statement as you learn more about the system, but it is important to capture your understanding of the Customer's NEED now. Dr. Burge's tool "18 Word Statement" is a great resource when developing your Needs Statement.

Action 1-B: Define Wider System

While defining the overall NEED in Action 1-A, you want to remember you are working with systems, and systems never act alone. Systems are part of a hierarchy, and they receive their purpose from the wider system. The result (NEED being satisfied) is created by the system and not just the element you are designing (recall your sound system). Systems interact with other elements, and these interactions must be identified and understood so that the element you are designing can properly fulfill its role in the entire system and unintended consequences can be avoided.

As you explore the problem situation in Action 1-A, a powerful tool to help you understand your system in the context of the wider system is a context diagram. A context diagram provides a quick snapshot of your system's purpose and the elements with which your system will interact. In systems terms, everything in your context diagram might either influence your system or your system might influence it. As you work through Action 1-A, you will create a context diagram from the view of the wider system and the day to day operations. These diagrams are described below.

**Context Diagram – Wider System View**

When we were discussing systems on page 23, you saw an example of a context diagram from the wider systems view for the test stand example introduced on page 5. This diagram is shown again in Figure 36.

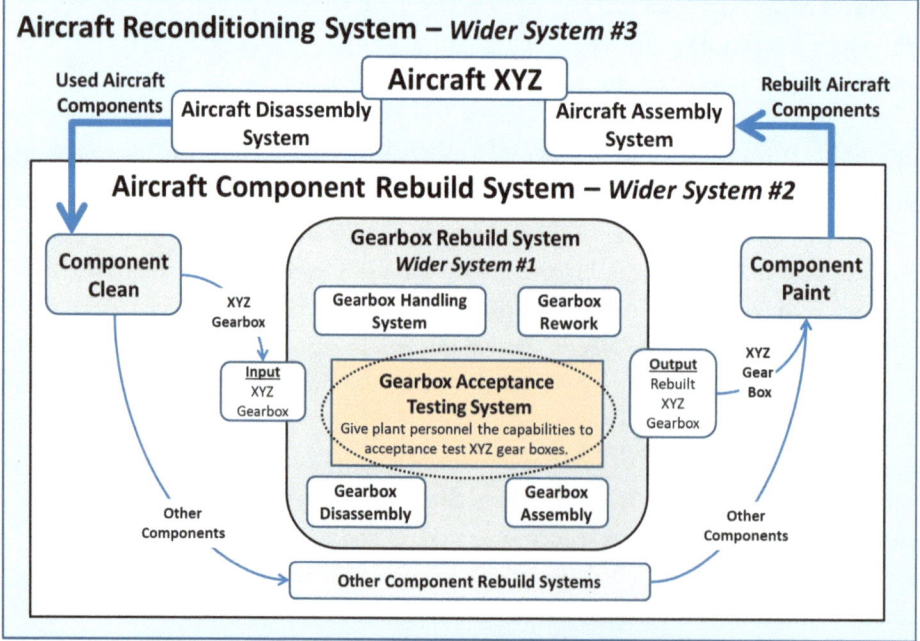

Figure 36: Context Diagram - Wider System View

This wider systems context diagram steps back and takes a big picture look at your system and its wider systems. Notice Figure 36 shows three levels of wider systems – Gearbox Rebuild System, Aircraft Component Rebuild System, and Aircraft Reconditioning System. When initially exploring your problem situation, I recommend identifying at least two levels of wider systems.

Review Figure 36 and notice some of the ways it can help you.

- It is very easy to visualize the big picture of the problem situation. Components are taken off of aircraft XYZ, rebuilt in the Aircraft Component Rebuild System, and then put back on the aircraft. Within the Aircraft Rebuild System, components are cleaned, rebuilt in a Rebuild System, and painted. It is so easy to get lost in the details during a design project. A quick review of this diagram brings the big picture back into view.

- It is easy to see the system elements that are removed from the immediate context of your system. Your system can often have significant impacts on elements in the wider system removed from its immediate context. These elements are easy to forget, and the context diagram from the wider systems view helps you keep in mind how your system may influence or be influenced by these elements. Your goal is always to design your system such that it improves the wider system. Reviewing the wider system context diagram is a quick way to review elements that may be impacted by a design decision within your system so you can optimize the whole system. The example in the next chapter shows how keeping the wider system in mind has such an impact that it turned a cost only project into a cost savings project.

- It is very easy to see where your system fits in. Your system is an element in the Gear Box Rebuild System which also includes Gearbox Disassembly, Gearbox Assembly, Gearbox Handling, and Gearbox Rework. These elements in your immediate context will be studied in detail in the context diagram from the operational view. During detailed design, it can be easy to be lost in the details and forget about the interactions your system has with these elements in your immediate context. A quick review of this diagram brings them back into your view.

- It is easy to see your system's purpose. Your system's purpose is to provide plant personnel with acceptance testing capabilities for XYZ gearboxes XYZ (Needs Statement given in the middle of the diagram).

- It is easy to see your system's boundary shown with a dotted line. The only element you are responsible for and can change is your system.

**Context Diagram – Operational View**

In addition to the big picture perspective given by the wider systems view, you will also focus on the immediate context of your system and its day to day operations to create a context diagram from the operational perspective. In this diagram, you dig into the details of the interactions your system has with the elements in its immediate context. The Operational view for the test stand example can be seen in Figure 37.

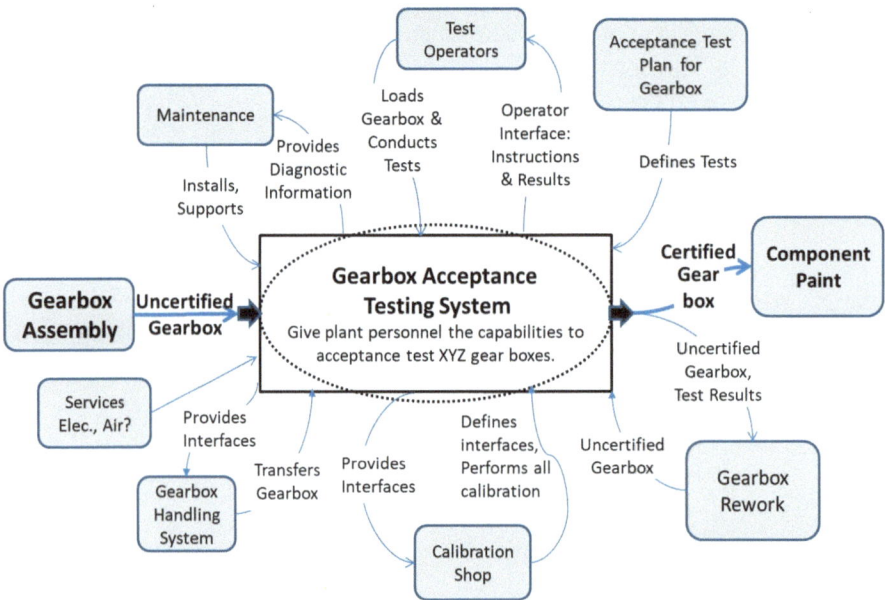

Figure 37: Context Diagram - Operational View

Initially, the context diagram from the operational view will help you define interfaces and the needs generated by the interactions your system has with other elements. It will also help you understand the elements you can and cannot change (system boundary noted with dotted line). As you progress through each stage of design, a quick review of the context diagram will help you maintain a perspective of the whole system.

Notice what you learn about the Gearbox Acceptance Testing System from the context diagram in Figure 37.

- The system to be designed is a device to perform acceptance testing. Its main purpose is to provide plant personnel with acceptance testing capabilities for XYZ gearboxes XYZ (Needs Statement given in the middle of the diagram).

- The only element you are responsible for and can change is your system (system boundary shown with dotted line).

- Most of the system requirements should be found in the acceptance test plan.

- The gearbox will be coming from the Gearbox Assembly. This element, as well as Gearbox Handling, will need to be consulted to understand how the gearbox will arrive and how you will interface with the handling system.

- The gearbox and certification status will go to the Gearbox Rework station. This station should be consulted when determining the content and format for the certification results.

- Your system will have to interface with plant services such as air and electric. You must consider what is available when performing detailed design.

- The company already performs some type of calibration (there is a calibration shop). You will want to talk with the shop since your device will likely have components that will need to be calibrated. You would not want to pick a device for your design if one that already exists in the plant would work. Picking a new one would mean a new item to stock, a new calibration procedure to write, and a new operating procedure for technicians to learn. Remember, you are working with systems, and there are always many ramifications to every design decision - even ones that seem small and insignificant.

- Completed gearboxes will go to Component Paint.

I find that at first context diagrams can be a little challenging, so I will step you through creating the two context diagrams using the MixItUp case study as an illustration. You should follow the steps for your system and create the two context diagrams. Remember, you will not create these diagrams in one sitting. You are bouncing back and forth between Action 1-A and 1-B. As you learn about the system, you will add to your context diagrams. Working on the context diagrams will generate questions and drive you back to learn more. This back and forth will continue until you have all of the questions in Action A-1 answered and you have the two context diagrams from Action1-B complete.

I highly recommend reading Dr. Burge's tool "Context Diagram." It is very insightful, and my steps below have been adapted directly from his article.

### Steps for your Context Diagram – Operational View

1. Pick a view for the Context Diagram. Your view is the day to day operations. You are looking at your system as a black box operating in the problem situation.

2. Draw a circle and write 1) the name of the system to be designed and 2) the Needs Statement. Figure 38

3. Draw a square for each external entity (element) in which the system interacts. The elements can be a tangible object or person or a mental construct such as the Quality System or Maintenance. Figure 38

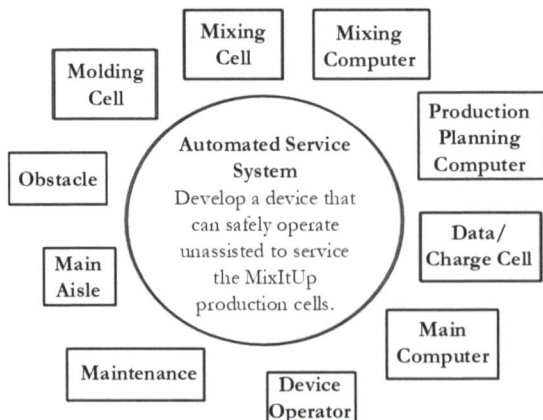

Figure 38: MixItUp Context Diagram - Operational View Steps 1 - 3

4. Determine the key interaction(s) between each element and the system and show these as a labeled arrow. Figure 39

   When identifying interactions, you are identifying WHAT happens between the two elements. The details of HOW it happens will be defined in Step 2 and Step 3. Consider a few examples from the MixItUp example in Figure 39.

   - What happens between your system and an Obstacle? The system detects the existence of an obstacle. The system responds appropriately to the obstacle.
   - What happens between your system and the Mixing Computer? You system requests the number of packs needed from the Mixing Computer. The Mixing Computer sends the requested data. The Mixing Computer also receives orders from Production Planning.
   - What happens between your system and the Mixing Cell? Your system must identify and position its self in the cell. Your system must deliver packets to the cell.

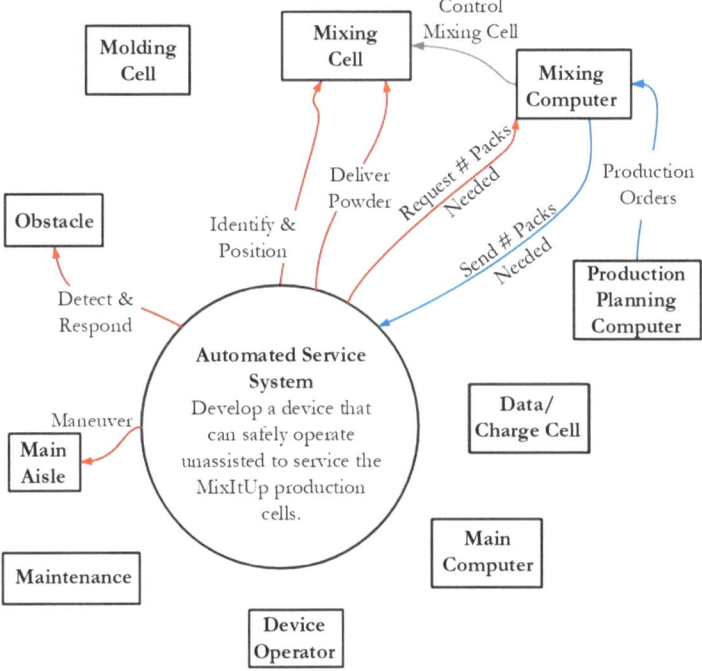

Figure 39: MixItUp Context Diagram – Operational View Step 4

Developing the context diagram from the operational view helps you to look in detail at all of the elements in your immediate context and understand the interactions with your system. It also helps you not to get involved in the details of these interactions too early by forcing you to identify the WHAT in the interactions. For example, consider "Send # Packs Needed" between your system and the Mixing Computer. When studying the system, you find that there is a 5 step handshaking protocol with information going between the computers, a specialized encryption process, and ten other steps. All of these details are important, and you will bring them out during Step 2 and Step 3, but right now, you want to identify the purpose of all these details clearly. You want the WHAT which is to "Send # Packs Needed." If security had actually been that important, then I would have put "Securely Send # Packs Needed" in the diagram.

The full context diagram for MixItUp can be seen in Figure 40.

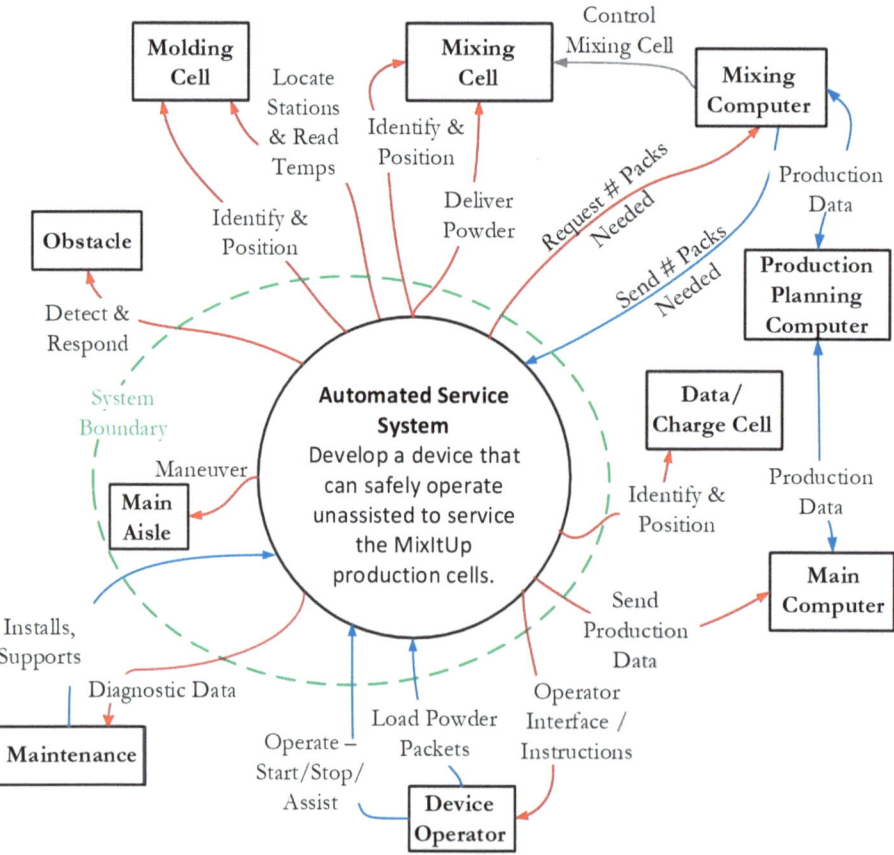

Figure 40: MixItUp Context Diagram - Wider

**Steps for your Context Diagram – Wider Systems View**

To develop the Wider Systems View Context Diagram for your system, follow the steps below.

1. Look at your operational view context diagram and identify the wider system which contains your system. Figure 41

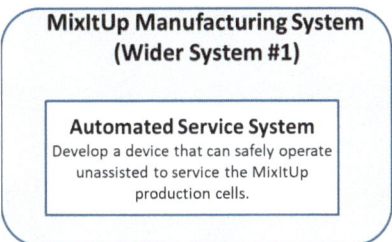

Figure 41: MixItUp Context Diagram - Wider System View Step 1

2. Identify the main inputs and outputs of the wider system. Figure 42

Figure 42: MixItUp Context Diagram Wider System View Step 2

3. Study the wider system and add the key elements necessary to transform the inputs into outputs. Most of these elements will be in your operational view context diagram. Figure 43

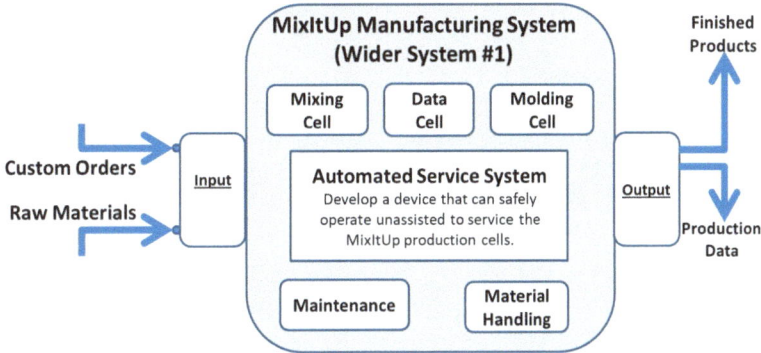

Figure 43: MixItUp Context Diagram - Wider System View Step 3

4. Consider your system as wider system #1 and repeat steps one through three to identify and diagram wider system #2. Wider system #2 is the wider system that contains wider system #1. Figure 44

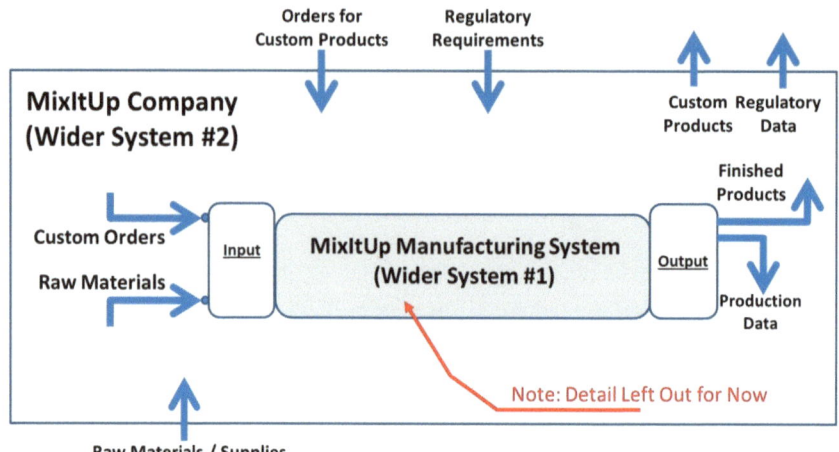

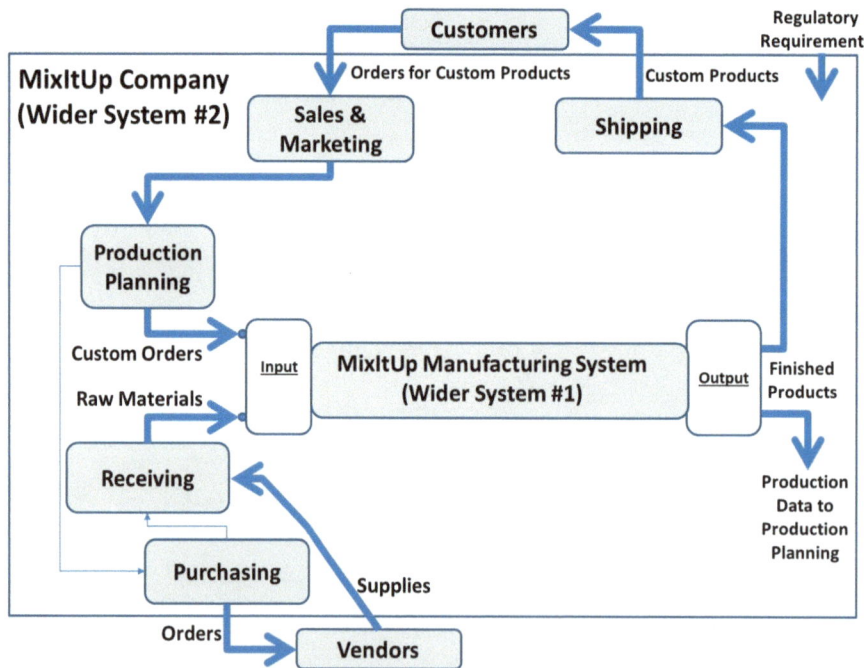

Figure 45: MixItUp Context Diagram - Wider Systems #2: Steps 1-3

The final wider system view context diagram for MixItUp is shown in Figure 47

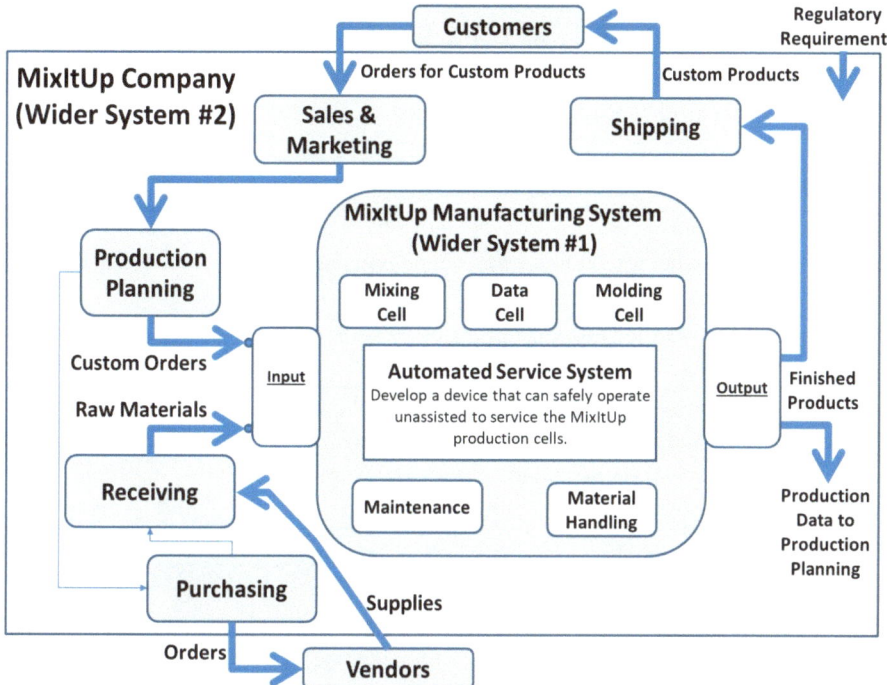

Figure 47: MixItUp Context Diagram - Wider System View

Action 1-C: Define Stakeholders

After understanding the problem context, you will identify the key people responsible for and influenced by the system. This is done by developing the Stakeholder Table as shown in Figure 48.

The Stakeholder Table is a versatile tool that can assist as you manage your project, identify needs, and implement your solution.

When identifying needs, the Stakeholder Table helps you determine where to go for different types of information and shows critical dynamics you must consider. Each group in the Stakeholder Table represents a unique perspective of the system, so during Step 2, you will explore each perspective to identify system needs.

Good communication is critical for any design project. By definition, the Stakeholder Table identifies all of the people that have a "stake" in your system. It contains the people that will be interested in the changes your system will bring, and you will want to keep these people informed. Good project managers develop a communications plan (what, when, how) for how they will keep each group in the Stakeholder Table informed.

Design projects always depend on people for a successful implementation. The Stakeholder Table is useful in identifying what is required from each group for success. It is also helpful to identify training needs for each group.

| NAME | CONCERNS | INVOLVEMENT |
|---|---|---|
| Plant Manager | Wants to make sure AGV does everything needed with at lower production cost and production delays | On Development Team |
| HR Manager | Concerned about the safety of everyone in plant and impact to workers | On Development Team |
| Maintenance Manager | Concerned about how his people will service it and interference when reconfiguring cells. | On Development Team |
| Quality Manager | Excited about getting reliable data. Wants to make sure it is collected at required intervals. | None – currently out of the country |
| Project Engineer | Responsible for the success of the project. | Main Point of Contact |
| Operators | Do not see how something automatic can do what they are doing now. | Available for questions |
| Other People in Plant | Unaware of project | None – Do not contact |
| Maintenance Workers | Kind of excited about having something new to work with. | Available for questions |

Figure 48: Example Stakeholder Table

Create a Stakeholder Table for your system by following the steps below.

- Work with the customer and identify everyone that has a stake in your system. Dr. Burge's tool "Stakeholder Influence Map" is helpful in identifying and organizing stakeholders.

- Define the column headings for your Stakeholder Table by identifying the information you want to capture for each stakeholder. Some example headings are shown in Figure 48. Others include job title, role, and knowledge area.

Action 1-D: Define Top-Level Functions

When people talk about design, you typically hear them mention form and function. The form is what the item looks like. The function is what the item does. The function is what makes an item useful. A car can have great form, but it is not useful if it does not perform necessary functions such as transport passengers, accelerate, and stop. Functions are statements of WHAT something must do and are independent of

HOW it will be accomplished. The high-level functions for a toaster may be Support Bread, Heat, Control Heat, and Keep User Safe. Depending on the market focus for the toaster, it might have an additional function of Interface with Whole House Controller. Notice that each function expresses WHAT must be done and not HOW. Before you can design something or even define the details of what must be designed, it is critical you have a clear picture of what your device must do. To understand what your device must do, you will define the high-level functions that are essential for your system to satisfy the customer's NEED and express them in the High-Level Functions Outline. An example High-Level Functions Outline can be seen for the case study in Figure 49. A diagram form of the outline is shown in Figure 50.

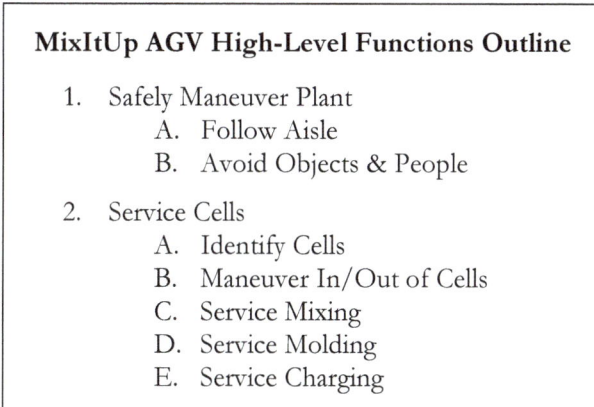

Figure 49: Example High-Level Functions

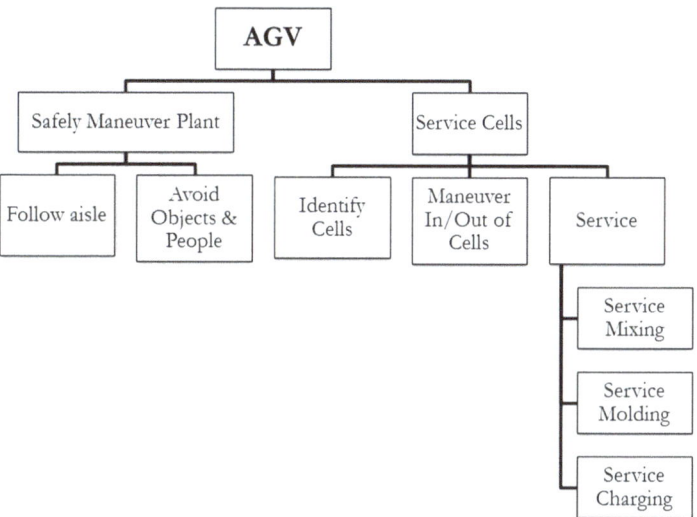

Figure 50: Example High Level Functions in Diagram

Engineering Design: A Systems Perspective - Page 73

Identifying the high-level functions helps you to understand the task in front of you better and exposes the essential needs that must be satisfied. It also provides the beginning of an outline where you will organize all of the facts learned about the system when discovering needs.

Create a High-Level Functions Outline for your system by following the steps below.

1. Review your Needs Statement and what you have learned about the problem during Action 1-1: Define the Overall NEED.

2. Think of the system as a black box and brainstorm WHAT it must do operationally to meet the expectations of the customer. As you brainstorm, try and express what it has to do using big picture action words – Identify, Communicate, Carry, Store, and Process. Do not get all wrapped up in exact wording. This list is for you right now, so use words that make sense to you. If you are doing this activity as a group, you may want to brainstorm on sticky notes and place the functions on a board.

3. Study your list of functions and group similar items.

4. Develop a "title" that describes the essence of each group and arrange your list in outline form as shown above in Figure 49.

5. Look at your outline and ask, "If these functions are successfully completed, will the customer's NEED be satisfied?" If the answer is yes, you have your initial list of High-Level Functions. If the answer is no, then you should fill in the gaps.

At the end of Step 1, you will have a clear understanding of

- the customer's overall NEED (Need Statement - Figure 35),

- the wider system and how it interacts with the system you are designing (Context Diagrams - Figure 40 & Figure 47),

- the key players in and influenced by the system you are designing and their role in its design (Stakeholder Table - Figure 48), and

- the key functions your system must perform to satisfy the customer's NEED (High-Level Functions - Figure 49).

With your initial understanding of the problem, stop and ask yourself if you are developing the correct system. Will the requested system satisfy the customer's NEED? Will the system fill the gaps in their current situation? If you do not think the requested system will satisfy the customer's NEED, you should have a conversation about this with the customer. If what they are asking for will satisfy their NEED, continue to Step 2.

**Step 2: Discover Needs (Wants and Needs)**

During Step 2, you will define specific needs by trying to understand the system in operational terms – from the customer's perspective. You will look at the system as a black box operating in the complex problem situation and identify everything that is necessary for it to satisfy the customer's NEED. If it *matters* to the *customer*, you want to capture it so you can include it in the design. Recall our discussion concerning wants and needs? Keep in mind that "*matters*" could mean something absolutely necessary (need) or just something that the customer would like to have (want). You will want to remember both types of needs, but when possible, you will want to clearly distinguish "like to have" items from "must have" items. Also, keep in mind that "*customer*" is not just the one that asked you to do the work. The customer you are concerned with at this point is everyone in the wider system that influences or is influenced by your system. This will mean physical entities that interact with your system as well as groups such as maintenance, calibration, purchasing, and human resources.

The two actions you will take during Step 2 can be seen in Figure 51, and a description of each action follows. Remember, when you are told to take specific action to understand needs, this does not mean you do it alone in isolation. You do it in partnership with the customer using all of the methods described in Figure 33.

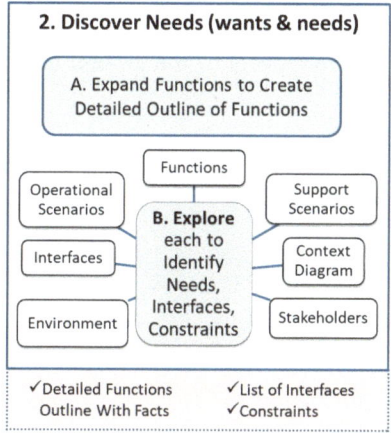

**Figure 51: Overview of Step 2**

Action 2-A: Create Detailed Outline of Functions

As you learn about your system, you need a way to organize all of the relevant information discovered. Since the system functions represent the essence of what your device will do, a detailed outline of the functions works well as a place to store and organize information. To create an outline of your system's functions, you will begin Step 2 by expanding the High-Level Functions Outline from Step 1 into a detailed outline of all the key functions the system must perform to satisfy the customer's NEED. This list is called the Detailed Functions Outline. Just like the High-Level Functions Outline developed in Step 1, the Detailed Functions Outline represents WHAT the system has to do. The Detailed Functions Outline for some of the functions in the MixItUp case study can be seen in Figure 52.

Compare functions in Figure 49 with the same function in Figure 52 and notice the detail added under the high-level functions. For example, notice the functions "Know Object is There" and "Respond Appropriately to Object" have been added under the function 1.B. "Avoid Objects & People." This additional detail represents WHAT must happen for the higher-level function to be accomplished and each WHAT is a functional need for your system. Your design must properly implement each function to satisfy your customer's basic need, so defining needs is largely uncovering the details of each function to make this happen.

When defining needs, you will use the Detailed Functions Outline to store and organize all of the information you learn while exploring the system. During detailed design, the outline can help generate ideas by providing a structure for brainstorming ways to accomplish each function (see page 107 for an example of this). As with the Context Diagram, periodically reviewing the Detailed Functions Outline is a good way to keep in perspective all of the aspects of the system that must be considered during detailed design.

### MixItUp Partial Detailed Functions Outline

1. **Safely Maneuver Plant**
    A. Follow Aisle
    B. Avoid Objects & People
        i. Know Object is There
        ii. Respond Appropriately to Object

2. **Service Cells**
    A. Identify Cell
    B. Maneuver In/Out of Cells

3. **Service Mixing**
    A. Accept and Carry Packets
        i. Accept Packets
            a. Know When Need Refilling
            b. Let Someone Know Need Refilling
            c. Allow packets to be loaded
        ii. Carry Packets
    B. Determine # Packets to Deliver
    C. Deliver Packets & Store Qty. Delivered

4. **Service Molding**
    A. Determine if Machine is Present
    B. Take & Store Temperature

5. **Service Data/Charging**
    A. Engage with Charging System
    B. Transfer Data to Main Computer

6. **Provide Operator interface**

7. **Misc./Constraints**
    A. Use NXT

Figure 52: Example Detailed Functions Outline

To expand your high-level functions into the Detailed Functions Outline, follow the steps below.

1. Review your Needs Statement and what you have learned about the problem during Action 1-A: Define the Overall NEED.

2. Review your High-Level Functions Outline.

3. Look at each high-level function in your outline and ask, "WHAT has to happen for this function to accomplish its purpose?" Write down the required functions under the higher level function.

4. Keep asking question 3 for each function at each level until there is no other WHAT. When specific solutions (HOWs) are all that come to mind when looking at a function, it is likely you have broken the function down far enough.

5. Step back and look at your detailed outline of functions and ask, "If these functions were completed, would the customer's NEED be satisfied?" If the answer is yes, you have your detailed functions. If the answer is no, you need to fill in the gaps. You will want to ask this question at each level in your outline.

6. Review your list and see if the names for each function all make sense to you. Adjust as necessary.

7. Add a "misc." item to your outline. You will use this category to capture facts about your system that do not fit anywhere else. You also may want to add a level for constraints and interfaces.

As you develop your Detailed Functions Outline, keep the following in mind.

- You are thinking about the system as a black box and identifying WHAT it must do and not HOW it will do it. Resist the temptation to jump to HOW too early.

- Be careful. The first few times you start talking to the customer about their needs, it is easy to get confused about WHAT and HOW. A customer's operational "HOW" is often still a "WHAT" for you as the designer. For example, a customer asks "HOW will an AGV deal with people in the way?" They are not asking technically how will you identify and deal with people in the path. They are asking operationally "WHAT will the AGV do in my plant when it sees a person in the way?" They want to know if it will stop, go around, set off an alarm, page someone, or take some other action. WHAT the AGV will do depends on the needs of their problem situation, and this is exactly what you are trying to define right now.

- Inevitably, there will be some HOWs discussed during this phase. By this point, you should have the message of how critical it is not to jump to HOW too early. It is your natural tendency as an engineer to focus on HOW instead of WHAT, so you must practice the discipline of WHAT before HOW. However, it is impossible to ignore completely how something will be accomplished when defining needs. The definition of some needs will require a general discussion of how the need will be implemented, but you are not committing to a solution. You should be on high alert any time you hear HOWs at this phase because every HOW you include as a need places restrictions on you when you go to design the system.

- Try and express what your system has to do using big picture action words such as identity, communicate, carry, store, and process.

- Do not get all wrapped up in exact wording. This list is for you right now, so use words that make sense to you. You can always go back and reword phrases as needed.

- Keep in mind that the Detailed Functions Outline you create is a rough draft. As you progress through understanding the system, you will discover other functions and update this outline.

Action 2-B: Explore System to Identify Needs, Interfaces, and Constraints

Now that you have the Detailed Functions Outline to organize what you learn about the system, it is time to explore the system from various perspectives to uncover all of the needs. Of course, you will view the system from an operational perspective, but you also must consider all other life-cycle demands that will be placed on the system during installation, support, and disposal. You want to capture anything that does (or should) matter to the customer, and you will do this by viewing the system from all of the perspectives shown in Figure 53.

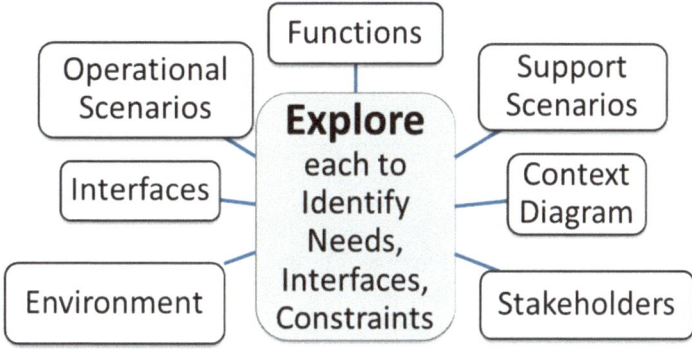

Figure 53: Exploring Needs from Different Perspectives

Considering each of the perspectives shown in Figure 53 is how you make the system "fit like a glove" when implemented. Together, the perspectives allow you to identify all of the details that matter to the customer.

When you view the system from a particular perspective, you will ask various questions. Each question you ask may or may not reveal information about needs. When it does, use the Detailed Functions Outline and record either the details you learn about a need or the questions that must be answered to define the need completely. When something does not relate directly to one of the functions in your outline, you will put it under the miscellaneous outline heading. As you learn more about the system, you may discover that some key functions are missing from your outline. This is normal, and you should add functions to your Detailed Functions Outline as you discover them.

The questions below will help uncover needs when viewing the system from any of the perspectives.

**GENERAL QUESTIONS TO EXPLORE NEEDS**
1. What is the wider system and what is its purpose?
2. What needs to be done?
3. How well does it need to be done?
4. Where does it need to be done?
5. Under what circumstances does it need to be done?
6. What/Who will cause it to do it?
7. For how long does it need to do it?
8. Under what conditions must it do it?
9. Are there any standards (company, local, federal) that must be followed?
10. Are there any relevant life-cycle issues to consider?
11. Is there anything special about how it must do it (i.e. cost, schedule, or technology)?

Figure 54: Questions to Explore Needs

A partial example of a Detailed Functions Outline with some of the details generated during Action 2-B can be seen in Figure 55 for the case study.

# MixItUp Detailed Functions Outline With Facts/Questions

1. **Safely Maneuver Plant**
   A. **Follow Aisle**
      i. There is a 9" clear aisle, and the AGV must stay inside of the aisle
      ii. Can mark the aisle
      iii. Must be able to go over 3/16" cords in the aisle
      iv. Would like AGV to go clockwise
      v. AGV will be stored at charging station
      vi. *How fast must AGV make a loop?*
   
   B. **Avoid Objects & People**
      i. **Know Object is There**
         a. Object is something > 3/16" high x 1" wide
         b. People cross the plant floor wherever they want.
      ii. **Respond Appropriately to Object**
         a. Do not attempt to go around objects
   
   C. **Respond to Getting "Lost"**

2. **Service Cells**
   A. **Identify Cell**
   B. **Maneuver In/Out of Cells**

3. **Service Mixing**
   A. **Accept and Carry Packets**
      i. **Accept Packets**
         a. **Know When Need Refilling**
         b. **Let Someone Know Need Refilling**
            1) Like to have a pager / No loud noises
         c. **Allow packets to be loaded**
            1) Must have a visual way to tell when full
            2) Like to have packets loaded from ground level
      ii. **Carry Packets**
         a. *What is the size of a packet?*
         b. *How many packets do we need to carry?*
   
   B. **Determine # Packets to Deliver**
      i. Protocol to interface with mixing computer is in station manual
   
   C. **Deliver Packets & Store Qty. Delivered**
      i. Cell cycle time for mixing 1 batch is always 6 minutes
      ii. Must ensure machine never runs out of packets
      iii. Maximum number of 1 type powder in a batch is 4
      iv. *How quickly can packets be delivered?*

Figure 55: Example Detailed Functions Outline with Facts

4. **Service Molding**
   A. **Determine if Machine is Present**
      i. There can be up to three machines
   B. **Take & Store Temperature**
      i. Must take temperature within ± 1 degree Fahrenheit
      ii. The temperature must be taken within ± .5 inches of the center of the machine to satisfy industry regulations
      iii. The Temperature must be taken every 15 minutes to satisfy industry regulations

5. **Service Data/Charging**
   A. **Engage with Charging System**
   B. **Transfer Data to Main Computer**

6. **Misc./Constraints**
   A. Use NXT
   B. Must be able to be transported by company forklift
   C. Must be able to communicate using Bluetooth (communications with Mixing and Main computers)

7. **Interfaces**
   A. **Physical Interfaces**
      i. AGV and Each Cell – Mixing, Molding, Data/Charge
      ii. AGV and Operator
      iii. AGV and Floor/Aisle
   B. **Logical Interfaces**
      i. AGV and Mixing Computer – see station manual
      ii. AGV and Main Computer – see company documentation

Figure 55: Example Detailed Functions Outline with Facts (Continued)

Notice the following in Figure 55.

- Item 1.C is an example of a new function that was discovered and added to the outline.
- Items 1.A.i-iv are examples of facts that have been discovered.
- Item 1.A.v is an example of noting an item that is desired but not necessary.
- Item 1.A.vi is an example of a question that must be resolved.
- Item 6.B is an example of using the "Miscellaneous" heading. You discovered that the company forklift would transport your system and this need did not fit under any of the functions.

To explore needs for your system, follow the instructions given below for each perspective.

**Functions**

You will explore each function in your Detailed Functions Outline using the questions shown in Figure 54. It will be very easy to assume you understand what is needed for a function, but slow down and go through each question for each function. It will not take that much time, and you will be surprised what you learn when you do. For example, consider some of the questions when looking at 4.B "Take Temperature" in Figure 52.

> 1. What is the wider system and what is its purpose?
>> The wider system is the MixItUp Manufacturing System. A review of the Wider System Context Diagram (Figure 47) should alert you that there may be specific regulatory requirements that relate to the temperature. You would need to go and research some to find out about these regulations.
>
> 2. What needs to be done?
>> This one is pretty easy; you need to take a temperature. This is the point where you may be tempted to think "this is too easy" and move on to the next function. Do not.
>
> 3. How well does it need to be done?
>> Don't skip this question. It is asking how accurate the temperature must be. You discover the temperature must be measured within ± 1 degree Fahrenheit. This is an important fact that will drive your choice of devices to measure temperature.
>
> 4. Where does it need to be done?
>> Again, this seems like a dumb question until you explore it a little more. You discover that the industry regulations require the temperature to be taken within ± .5 inches of the center of the machine's heating element. I am glad you took the time to ask!
>
> 5. Under what circumstances does it need to be done?
>> You discover that the temperature must be taken every 15 minutes to satisfy industry regulations.
>
> 6. What/Who will cause it to do it?
>> Your system must do it automatically.
>
> 7. For how long does it need to do it?
>> Nothing discovered here.

For a simple function like "Take Temperature," the questions in Figure 54 sure uncovered many facts important to satisfying the customer's NEED. Notice the items below were captured under 4.B in the example Detailed Functions Outline with Facts shown in Figure 55.

- Must take temperature within ± 1 degree Fahrenheit.

- The temperature must be taken within ± .5 inches of the center of the machine's heating element to satisfy industry regulations.

- The temperature must be taken every 15 minutes to satisfy industry regulations.

To explore needs for your system from the functions perspective, use the questions in Figure 54 and explore each function in your Detailed Functions Outline. Remember to take your time and be thorough. Record what you learn in your Detailed Functions Outline with Facts.

**Operational scenarios**

Since you have explored all of the system functions, you have learned many things about the system and the customer's needs. Now you will envision the system in operation on a day-to-day basis and try to identify everything that will be asked of it and everything that will happen to it. Each different case you think of is an operational scenario, and each scenario may uncover new needs for the system. Developing scenarios will also require you to define WHAT your system will do in certain cases. Some example scenarios for the case study are given below.

1. Normal Operation. The AGV leaves the charging station and maneuvers the aisle in a clockwise direction (notice 1.A.iv in Figure 55). It services all cells encountered and returns to the charging station where it transfers its data to the main computer. If packets are needed, it alerts an operator and waits for confirmation that packets have been loaded before continuing.

2. Obstacle Encountered. While following the aisle, the AGV encounters an obstacle. The AGV stops. If the obstacle moves within 10 seconds, the AGV will continue. If the obstacle does not move, it will alert an operator and wait for the obstacle to be moved. The AGV waits for operator input before continuing.

3. ERROR. AGV stops operating. Maintenance will be called, and if necessary, the AGV will be taken to the maintenance shop for repair.

Notice that operational scenarios 1 and 2 define WHAT your system will do as events occur. This definition is one of the powers of operational scenarios. They force you to work through the operational details of your system and define its operation. This definition is an excellent communications tool for the customer. It allows the customer to "see" your system in operation, so they know if they like WHAT the system does. For example, the customer may not like the fact that the operator has to give input to the AGV to restart after an obstacle is removed (scenario 2). Instead, they want the AGV to continue automatically when the obstacle is removed. The scenario just helped you define a need that will allow your system to "fit like a glove" when implemented.

Operational scenario 3 is how you discovered the AGV needs to be able to be transported by the company forklift (see 6.B in Figure 55).

You will use operational scenarios during every stage of design, and the detail contained within the scenario will increase as you progress through the design stages.

To explore needs for your system from the operational scenarios perspective, do the following.

- Make a list of all the likely scenarios for your system from an operational perspective. Don't forget to consider the various modes of operation your system may have such as automatic, manual, calibration, and maintenance.

- For each scenario, explore what your system needs to be able to do to accommodate the scenario properly.

- When necessary (like Scenarios 1 and 2 above), define WHAT your system will do in each scenario. At first, this definition will be very general, and then it will become more specific as your progress through Conceptual Design and Detailed Design.

**Support scenarios**

Support scenarios are just like operational scenarios except you are thinking about the actions required to support the system during its operational life. You will think about the system in use and review all of the tasks required to support it such as calibration and maintenance. Each support task may generate needs. For example, consider our test equipment example on page 5. When reviewing support scenarios, you would have identified the need for calibration and learned that all calibration is done off-site.

To explore needs for your system from the support scenarios perspective, do the following.

- Make a list of all the likely scenarios for your system from the support perspective.

- For each scenario, explore what your system needs to be able to do to accommodate the scenario properly.

- When necessary, define WHAT your system will do in each scenario. For example, when it is time for sensor calibration, the system will display an alert that will not go away until calibration is performed.

**Environment**

Explore all of the environments in which the system must operate and be stored. Do any of the environments create special needs for your system? Don't forget to consider environmental factors such as dust, rain, humidity, temperature, and electromagnetic interference.

**Context Diagram**

The Context Diagram is very important because it shows you all of the key entities with which your system will interact. In systems terms, your Context Diagram shows you the components of your wider system. Your system may influence and be influenced by each of these entities in your Context Diagram. Together, your system and these entities allow the wider system to fulfill its purpose. Your goal as a designer is to design your system so that it works with the other components in your Context Diagram to best meet the purpose of the wider system. You do not want your system to be the "best." You want your system to make the wider system the "best." The only way to accomplish this is to fully understand the interactions between your system and the entities in its Context Diagram.

To explore needs for your system from the context diagram perspective, do the following.

- Review your Context Diagram and explore the relationship between your system and each element in the Context Diagram.

- For each element, consider what the element wants from your system. What needs does this generate?

- For each element, consider how that element impacts your system. What needs does this generate?

**Interfaces**

Most system problems occur at interface points, so identifying and defining every interface for your system is critical. At the end of defining requirements, you will have an interface document that defines each system interface in detail, but at this point, you are just identifying all of the interfaces and exploring needs that arise from the interface. For example, when exploring needs from the interface perspective for the case study, you would have identified all of the interfaces listed under item 7. "Interfaces" in Figure 55. Exploring needs based on these interfaces would have caused you to identify item 6.C. "Must be able to communicate using Bluetooth" since the mixing computer and main computer both use Bluetooth for communications.

To explore needs for your system from the interfaces perspective, do the following.

- Make a list of everything with which your device must interface. You have already identified most items when exploring the context diagram. Add the items to your Detailed Functions Outline with Facts.

- Consider each interface and identify any needs generated by the interface.

**Stakeholders**

Each stakeholder represents a different role in the complex problem situation, and each role will view the system differently. For example, a maintenance worker will view your system very differently from someone that operates it each day. They both will view it differently from the plant manager, and it is unlikely that any of them have a complete view of the full system. Each role interacts with the system in different ways and needs different things from it. By exploring these different roles, you can gain a more comprehensive understanding of what is needed for your entire system.

To explore needs from the stakeholder perspective, review your Stakeholder Table and explore what each stakeholder wants out of your system. In addition to the questions in Figure 54, consider 1) when they want the system to be ready and 2) what other deliverables they want such as training, operations manuals, and spare parts recommendations.

A few tips for working with stakeholders are given below.

- You may not be able to talk with all of the stakeholders, but at least think through each of the questions from their point of view.

- Your system will not solve all of the problems of every stakeholder. Keep the focus on what the purpose of your system is, and explore needs that relate to this purpose.

- Stakeholders will often tell you exactly how the new system should be. Beware. Each stakeholder is usually only seeing the system from their point of view. Even if they are correct in how it should be, this is only how it should be to optimize their role. Your job is to understand the needs from their role and balance them with needs from all of the other roles to create the best overall system.

- Be very careful not to promise (or even imply) that the new system will have certain features that a particular stakeholder wants. You do not know what the system capabilities will be yet, and you do not want to set false expectations. The best way to have a failed implementation of your new system is to have people expecting something that they are not going to get.

At this point, you have explored the system from each perspective shown in Figure 53 and recorded what you have learned in your Detailed Functions Outline with Facts. You will leave Step 2 with (all shown in Figure 55)

- a Detailed Functions Outline that contains information about all of the functional and non-functional needs of the system,

- an understanding of the system interfaces, and

- a list of system constraints.

**Step 3: Clarify Needs**

During Step 3, you are clarifying the needs. Upon completion of Step 3, you will understand everything necessary to get started in the right direction to design a system that will satisfy the customer's NEED. The three actions you will take during Step 3 can be seen in Figure 57, and a description of each action follows.

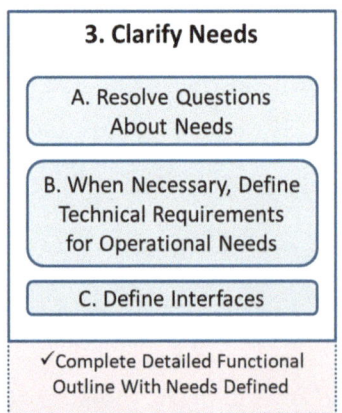

Figure 56: Overview of Step 3

Action 3-A: Resolve Questions about Needs

To clarify the needs, you will answer all questions generated during Step 2. The goal of clarification is to define every need with the details necessary to create a design to satisfy it. This may involve talking with the customer, performing research, or making a high-level architecture decision about your system's design. A few examples of clarification are given below.

> In Figure 55, notice the question "What is the size of a packet?" (3. A. ii. a)).
>> This question must be answered to design something to hold packets and can likely be resolved simply by talking with the customer.
>
> In Figure 55, notice the question "How quickly can packets be delivered?" (3. C. iv.).
>> You ask the customer for a time, and they refer you to the manual for the automated loading system for the mixing station. After some research, you learn that the mixing delivery system removes a dropped packet within two seconds of it hitting the floor. So, you cannot deliver packets any faster than two seconds. Your engineering judgment will have to lead you in setting the actual limit for your system (2.0 seconds, 2.1 seconds, 2.5 seconds, 3 seconds…).

Go through your Detailed Functions Outline with Facts and resolve any questions.

Action 3-B: Define Technical Requirements for Operational Needs

Sometimes you have a need defined operationally, but this definition does not provide enough detail for you as the designer. In these cases, you must take the operational need and further define it to achieve a technical requirement. Some examples follow.

> In Figure 55, you identified a need that your device must be able to be lifted by the customer's forklift (6. B.).
>> This is an appropriate operational need. It reflects exactly what the customer wants, but it is not defined well enough for design. For design, you will need to create technical statements to ensure you satisfy this customer NEED. For example, the device can weigh no more than X pounds (limit of company forklift), and the device must accommodate forks of size x by y. These details would be added to your Detailed Functions Outline with Facts.
>
> Figure 55 contains the question "How fast must AGV make a loop?" (1. A. vi.).
>> For this question, you talk with the customer and find that they do not care about the AGV speed. Their concern is that data from Molding is taken at least every 15 minutes (4. B. iii.) and that the mixing machine never runs out

of packets (#. C. ii.). Again, these are appropriate operational needs, but they are not specific enough to use as a design parameter for the AGV speed. You would have to study the operational characteristics of the Mixing station and determine the maximum time between deliveries to ensure Mixing never runs out of packets. Defining this time may require you to make a high-level design decision such as if your AGV will deliver a single batch of packets or multiple batches. Once the delivery frequency is calculated, you would have to determine the speed required for your AGV to deliver packets at this interval. Finally, you would have to ensure that this speed would allow you to collect data at Molding at least every 15 minutes. All of the details determined during this process would be added to your Detailed Functions Outline with Facts.

Review the needs in your Detailed Functions Outline with Facts and define technical requirements as necessary.

Action 3-C: Define Interfaces

Most system problems occur at interface points, so before leaving Step 3, you should review all of your system interfaces to ensure they each are defined as clearly as possible. As your design progresses, further details concerning the interfaces will surface, but at the end of Step 3, you want to record everything that can be known about them. For most systems, you will want to create an interface document that contains every interface and all of the details that define the interface.

To define interfaces for your system, do the following.

- Explore each interface identified in Action 2-B and define the interface as completely as possible.

- Create a draft interface document. The document should contain each interface and your current knowledge about the interface.

You will leave Step 3 with a completed Detailed Functions Outline with Facts. This outline contains your current knowledge of needs, and it is everything necessary to get started in the right direction to design a system that will satisfy the customer's NEED. However, this outline is not complete. Recall the analogy of a ship approaching port from the sea given on page 32. At the end of Step 3, the Detailed Functions Outline with Facts completely defines your current understanding of the needs, but you are still far from shore. This outline should be viewed as a working document. Throughout the design process, you will constantly be uncovering critical information, and you will want to update the Detailed Functions Outline with Facts with this new knowledge. This outline can even be used during detailed design to document specific design decisions.

**Step 4: Express Needs for Customer and Designers**

During step 4 you will work with the customer to determine the specific needs to be addressed by your system and then create the system requirements. The two actions you will take during Step 4 can be seen in Figure 58, and a description of each action follows.

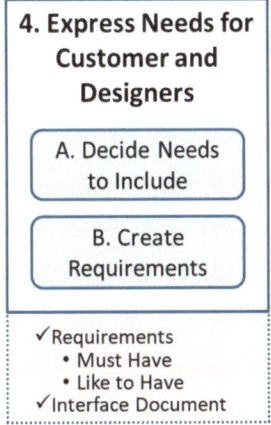

Figure 58: Overview of Step 4

Action 4-A: Decide Needs to Include

When talking about the success of a project, people often talk about the three-legged stool of success as shown in Figure 59.

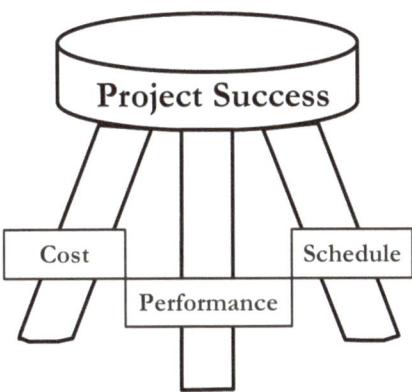

Figure 59: Tradeoffs in Design

The basic idea is pretty simple. In any specific project, cost, performance, and schedule are related, and you cannot change one without impacting one of the others. For example, consider building a house. After making all of the plans and setting a move-in date, you decide you want to increase the size of the kitchen (change in performance). This change definitely impacts cost, and it may impact your move-in date (schedule). If you cannot change the move-in date, then more resources will have to be allocated to get the extra work done in the same amount of time (impact to cost).

The idea illustrated in the three-legged stool also applies to design. Every design requirement specifies something your system will or will not do (Performance leg of the stool). Every design requirement also may impact the design's cost and or schedule (design cost/time as well as fabrication cost/time).

The Detailed Functions Outline with Facts contains all of the wants and needs for your system. Now you must use your knowledge and experience to work with the customer and determine which wants and needs you can realistically satisfy with your design given the current limitations on technology, available budget (cost), and time until delivery (schedule). You are going to have to lead the customer through making some tradeoffs. For example, the MixItUp maintenance manager is sure that he "needs" troubleshooting diagnostics built into the AGV. You can provide the requested diagnostics, but it would require moving to a hardware platform other than the Lego NXT. Recall that another "need" was to use the NXT. You discuss the ramifications of the requested diagnostics with the maintenance manager, and he realizes that he does not "need" diagnostics as much as he thought he did.

To determine the specific wants and needs to be addressed by your system, do the following.

- Review each item in the Detailed Functions Outline with Facts and identify items that you think may not be realistic based on current technology and limits on cost and schedule.

- Discuss the items in question with the customer and eliminate items as necessary. Help the customer understand the impact their requests have on cost, performance, and schedule. Remember, if an item is essential to satisfying the customer's NEED, you can modify it, but you can not eliminate it.

- Update your Detailed Functions Outline with Facts based on the items you decided will not be included in your design.

Action 4-B: Create Requirements

The Detailed Functions Outline with Facts now represents the requirements for your system. If you develop your design using all of the information contained in the outline, you should have a system that satisfies the customer's NEED.

Where you go from here depends on your development environment.

> If the design you are working on is your own project, you can use the Detailed Functions Outline with Facts to design the system.

> In some business environments, you will reformat your outlined needs list into a set of Functional Requirements for the system.

> In other situations, you will summarize all of the information from your requirements definition process into a document often called the Operational Concept Document.

> If you are in a formal systems engineering environment, you will turn your outlined list of needs into a set of formal requirements. These requirements typically will be presented in a Requirements Matrix that includes how each requirement will be verified. Some example requirements for the MixItUp case study are shown in Table 1. The requirements in Table 1 are organized around the top level functions from Figure 49. The right four columns indicate how you will show your completed product satisfies each requirement (**I**nspection, **A**nalysis, **D**emo, **T**est), and the number in the last column is a reference number for the test description. Some requirements reference the Interface Control Document (ICD) developed in Action 3-C.

## Table 1: Example Requirements for MixItUp AGV

| Item | Description | I | A | D | T |
|---|---|---|---|---|---|
| **1.0** | **Safely Maneuver Plant** | | | | |
| 1.1 | The Device shall operate with no human interaction except as noted in requirements for initial start/stop, packet loading, and trouble assistance. | | | | 1 |
| 1.2 | The Device shall navigate around the plant within a 9" wide aisle in a clockwise direction in accordance with ICD (Section 10.1). | | | | 1 |
| 1.3 | The Device shall service a cell each time a cell is encountered. | | | | 1 |
| 1.4 | The Device shall be stored at the Data cell when not in operation. | x | | | |
| 1.5 | The Device shall begin and end operation by input from an operator in accordance with ICD (Sections 10.8.1 10.8.2). | | | x | 1 |
| 1.6 | If Device loses its orientation, it shall stop and alert maintenance for assistance. | | | x | 2 |
| 1.7 | After a lost state, the Device shall resume operation after being placed back on the track and the resume button is pressed. | | | x | 2 |
| 1.8 | The Device shall be capable of driving over cords with a maximum diameter of 3/16." | | | x | 1 |
| 1.9 | The Device shall stop if any obstacle over 3/16" height from the floor is encountered. | | | x | 1 |
| 1.10 | After stopping for an object, the Device shall continue normal operation when the object is no longer detected. | | | x | 1 |
| 1.11 | When the Device has been stopped for an object for 10 seconds, the Device shall alert someone for assistance. | | | x | 1 |
| **2.0** | **Service Cells** | | | | |
| 2.1 | The Device shall detect cells located at any position on the inside perimeter of the aisle. | | | | 1 |
| 2.2 | The Device shall identify whether it has encountered a mixing cell, molding cell or Data cell. | | | | 1 |
| 2.3 | The Device shall automatically enter each cell. | | | | 1 |
| 2.4 | The Device shall physically interface with the molding cell in accordance with ICD (Sections 10.2 and 10.3). | x | | | |

Table 1: Example Requirements for MixItUp AGV (Continued)

| Item | Description | I | A | D | T |
|---|---|---|---|---|---|
| 2.5 | When the Device exits a cell, it shall continue along the path to the next cell. | | | | 1 |
| 3.0 | **Service the mixing workstation** | | | | |
| 4.0 | **Service the molding workstation** | | | | |
| 4.1 | The Device shall detect the presence of up to three molding stations arranged in accordance with ICD (Section 10.3). | | | | 3 |
| 4.2 | For each molding station present, the Device shall take the temperature value within ± .5" of the center of the machine's heating element. | | | | 3 |
| 4.3 | The Device shall take the temperature values within ± 1.5° F. | x | | | 3 |
| 4.4 | The Device shall store the temperature value (± 1.5° F) for each station present. | | | | 3 |
| 4.5 | The Device shall service the molding cell at least once in every 15-minute interval. | | | | 1 |
| 5.0 | **Service the Data workstation** | | | | |
| 6.0 | **Constraints** | | | | |
| 6.1 | The Device shall be operational by 12/14/2016. | | | x | |
| 6.2 | The Device shall use off the shelf NXT compatible components for all electronics. | x | | | |
| 6.3 | To accommodate being lifted by the company forklift, the Device shall not exceed 4kg. | | | x | |
| 6.4 | To accommodate being lifted by the company forklift, the Device shall provide fork guides 6 inches apart. | x | | | |
| 7.0 | **Interfaces (see Interface Control Document)** | | | | |
| 8.0 | **Like to Have** | | | | |
| 8.1 | Would like for packets to be able to be loaded from the ground level. | | | | |
| 8.2 | Would like for alerts from AGV to be given to a Bluetooth pager. | | | | |

Create your system-level requirements by formatting your Detailed Outline with Facts based on your development environment.

**Step 5: Validate Requirements**

During Step 5, you will update all of your requirement artifacts (documents and visuals), define design criteria, and ensure your understanding of the requirements are correct by having the customer review and agree to your requirements and acceptance criteria. The specific actions you will take during Step 5 can be seen in Figure 60, and a description of each action follows.

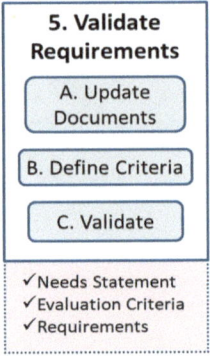

Figure 60: Overview of Step 5

Action 5-A: Update Documents

Go back to documents such as your Needs Statement, Context Diagram, Stakeholder Table, and High-Level Functions and update them as needed. You should have been updating these documents as your knowledge of the system increased, but it is good to review them once more now. You will refer to many of these document in the other states of design, so you want them to represent your best knowledge of the system.

Action 5-B: Define Criteria

During the design of your system, you will consider many options and make many decisions. Part of making your system "fit like a glove" is viewing the design and making decisions as the customer would. One way to do this is by working with the customer to identify Evaluation Criteria.

Think about purchasing a car and imagine you find two cars that both satisfy your basic transportation NEED. How do you select between the two choices? To choose, you evaluate the two cars based on what is important to you and select the one that performs best in the areas that are important to you. You and I may select different cars because different things are important to us. For example, I may be looking for how sporty the car looks, how well it handles on curves, and how fast it accelerates. You may be looking at fuel efficiency and safety. The things important to us that we use to select between two alternatives that satisfy our NEED are called Evaluation Criteria. You need to understand what is important to the customer so you can think like they would when making design choices. The Evaluation Criteria for MixItUp can be seen in Table 3. When making design decisions, the Evaluation Criteria will be used with a tool such as the Pugh Matrix (see Dr. Burge's description of this tool).

Table 3: Example Evaluation Criteria

| MixItUp Evaluation Criteria ||
|---|---|
| Weight | Criteria |
| 35% | Safety |
| 25% | Flexibility |
| 25% | Reliability |
| 15% | Cost |

To develop Evaluation Criteria for your problem, work with the customer and identify three to five criteria they want you to use when making design trade-offs. The Paired Comparison Method can be used with various stakeholders to develop relative weights for the criteria.

Action 5-C: Validate Requirements

Recall the stages of design shown in Figure 29 (page 50). The customer's NEED is expressed in the system-level requirements, and then the requirements drive all of your design efforts. Before you move to the other stages of design, you want to make sure the key stakeholders agree that your requirements accurately express their NEED. In a formal Systems Engineering environment, this would take place at a technical review called Systems Requirements Review. Regardless of the format for the review, you want your key stakeholders to review and sign off on the following.

- System-Level Requirements
- How you will verify that your system satisfies each requirement
- Evaluation Criteria
- Interface Document (definition of each external interface)
- Acceptance Criteria – what has to happen before the customer accepts the system and makes the final payment.

**Summary**

After completing the five steps described above, you will have thoroughly explored the problem situation and defined requirements which express exactly what is necessary to satisfy the customer's NEED. You know details will emerge as you progress through your design (ship approaching shore analogy), but you are now ready to move on to Conceptual Design. As shown in Figure 29, you will constantly review these requirements to ensure that every design decision considers all of the requirements necessary to satisfy the customer. Forgetting about even one requirement can allow you to pursue a design approach that becomes infeasible when the forgotten requirement resurfaces. To keep this from happening, keep the System-Level Requirements, Needs Statement, High-Level Functions, and Context Diagrams close to you and review them often.

# Chapter 5 - The Systems Perspective Complements Traditional Design

*The systems perspective gives you the tools to effectively understand the situation and define everything required to satisfy the customer's NEED. It reminds you to periodically stop and consider the wider system and the true purpose of your system. It also allows you to keep in mind the critical interactions between elements, so you properly accommodate these interactions in your detailed technical designs.*

*Your technical engineering design skills allow you to develop a product that satisfies the requirements.*

---

Chapter Two introduced you to the systems perspective, and Chapter Three gave you an overview of design. In Chapter Four, you learned how to define requirements for your specific design problem. This chapter uses a design example to show you how the systems perspective and your traditional technical design skills complement each other.

Recall the discussion earlier on what it is you really do as an engineer. Figure 2 is repeated in Figure 62 for reference. A customer with a vague problem within a complex ("Messy") problem situation comes to you with a NEED, and they want you to satisfy their NEED completely. They want you to study their complex problem situation, accurately understand the NEED, and develop a product (system) that, when implemented, will satisfy their NEED.

The systems perspective gives you the tools to effectively understand the situation and define everything required to satisfy the customer's NEED. It equips you to define requirements (top portion of Figure 62). Once you have a clear picture of what you need to design, your technical engineering design skills allow you to develop a product that satisfies the requirements (bottom portion of Figure 62).

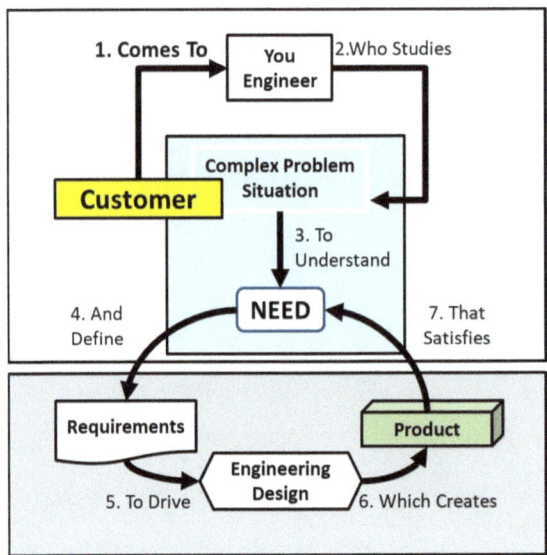

Figure 62: Essence of What We Do as Engineers

When performing detailed design, the systems perspective complements your traditional design skills. It helps you to remember that you are working with a system (recall the 4 Cardinal Rules for an Effective Design shown in Figure 31). It reminds you to periodically stop and consider the wider system and the true purpose of your system. It also allows you to keep in mind the critical interactions between elements so you can properly accommodate these interactions in your detailed technical designs.

To help you see how the systems perspective works with your traditional engineering design skills, I am going to put you on a design team and allow you to step through a real design problem.

A customer has come to you requesting a piece of material handling equipment. Specifically, they want you to design a dolly that will support, transport, and orient a gearbox as shown in Figure 63.

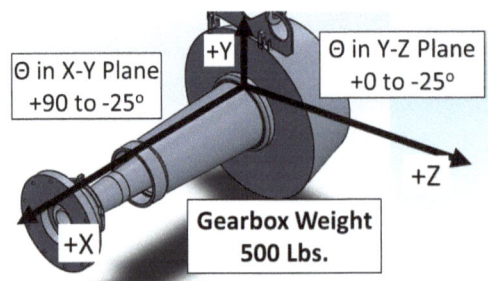

Figure 64: Gearbox Handling System - Initial Request

The company is the same company you have already been introduced to in the test stand example. They recondition military aircraft in support of an overall Air Defense System. The requested dolly is for the Gearbox Rebuild System where the test stand was an element. Recall that gearboxes are taken off the aircraft and then processed before arriving at the Gearbox Rebuild System. In the Gearbox Rebuild System, gearboxes are disassembled, rebuilt, and acceptance tested. The certified gearbox is then painted and reassembled onto the aircraft.

The customer has studied their situation and concluded that they need a different type of dolly to increase throughput inside of the Gearbox Rebuild System. The current device orients the gearbox as necessary for teardown and rebuild, but it is stationary. Gearboxes are transported to the Gearbox Rebuild System on another device and then transferred to the current device for teardown and rebuild. Rebuilt gearboxes are transferred again and transported to the test station. Their vision is a dolly that accepts the gearbox from the transport device and then accommodates all work and movement needed within the Gearbox Rebuild System.

The description above is typical of how your design problem will begin. The customer has a NEED and a general idea of how they want you to satisfy it. They come to you with some information, but many details that you will need to solve their problem are missing.

The following steps will give you a general idea of what you do from this point forward. Throughout the steps, you will see the systems perspective and your traditional engineering design skills working together to allow you to satisfy the customer's NEED.

**Requirements Definition.**

You discover needs and develop requirements by working through the five steps described in Chapter 4. You study the wider system, talk with the customer, talk with workers within the Gearbox Rebuild System, and study the other elements within the Gearbox Rebuild system. The customer asks you for a Dolly, but you practice the discipline of WHAT before HOW and resist jumping straight to HOW. Your initial Needs Statement is shown in Figure 65.

> Create a device to support, orient, transport, and allow the rebuild of the XYZ gearbox.

Figure 65: Gearbox Handling System - Needs Statement

You create Figure 66 to understand the wider system better, and you learn that the gearbox undergoes processing before (Component Clean) and after (Component Paint) the Gearbox Rebuild System.

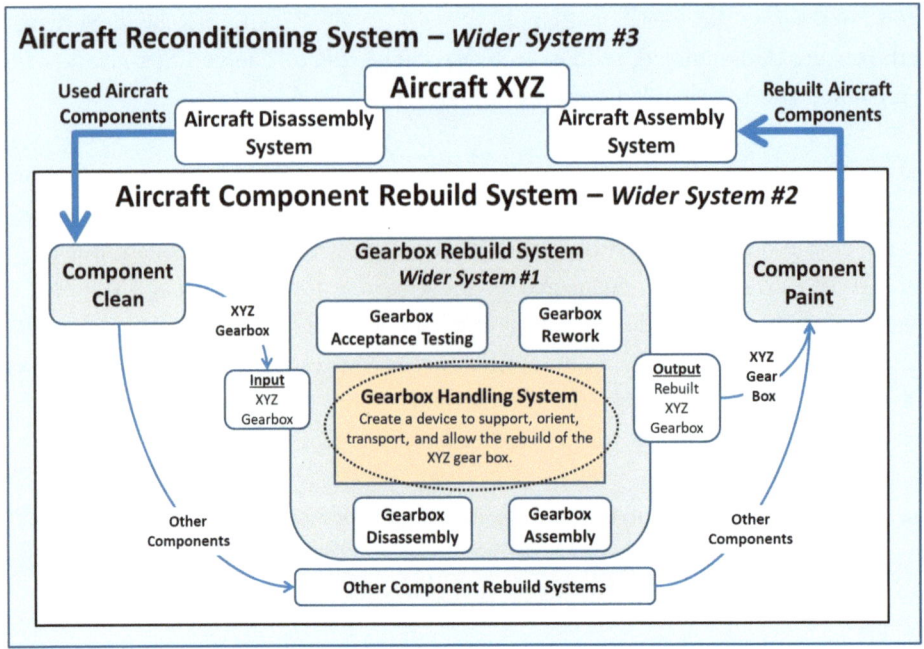

Figure 66: Gearbox Handling System - Wider System

When studying the wider system, you learn that full access to the area behind the gearbox mounting plate is needed to perform the work required at Component Clean and Component Paint. Currently, the gearbox is placed on a transport dolly from the aircraft and then removed and placed on a special dolly at Component Clean, Component Paint, and at the Gearbox Rebuild System. At each station, it is then put back on the transport dolly to move to the next station. Every transfer requires time and is an opportunity to damage the gearbox and injure the operator. Even though you are focused primarily on the Gearbox Rebuild System, your systems perspective causes you to consider your device in the context of allowing the wider system to achieve its purpose best. If your device could be designed to accommodate the access required at Component Clean and Component Paint, the gearbox could be placed on your device directly from the aircraft. It could then remain on your device while it travels to each process and then back to the aircraft. All transfers and opportunities for damage and injury would be eliminated. The customer's NEED to increase throughput would be met, and the chance of injury and damage to the gearbox would be reduced.

From the perspective of the Gearbox Rebuild System, adding a need that the device must allow access to the area behind the gearbox mounting plate is not a good thing. Access is not necessary for the Gearbox Rebuild System and making access a need increases the complexity and cost of the design.

Requiring the device to allow access is not good for the Gearbox Rebuild System, but your system perspective causes you to look past what is best for your element to what is best for the overall system (Aircraft Component Rebuild System) in satisfying the customer's NEED. You discuss the access situation with the customer and help them understand the access issue from the perspective of the overall system. Together, you determine that the overall savings in time and reduction in potential gearbox damage and operator injury more than offsets the increase in the cost of the device. In fact, the change will save money, so the project goes from a cost only project to a cost savings project. You agree to add a need that your device allows access to the area behind the mounting plate. The systems perspective has identified the need, and now your traditional engineering design skills will determine how to rotate a 500-pound gearbox without support in the center of rotation; Good luck!

As you continue to discover needs, you develop the operational view context diagram (Figure 67) which helps you understand the elements in which your system will directly interface.

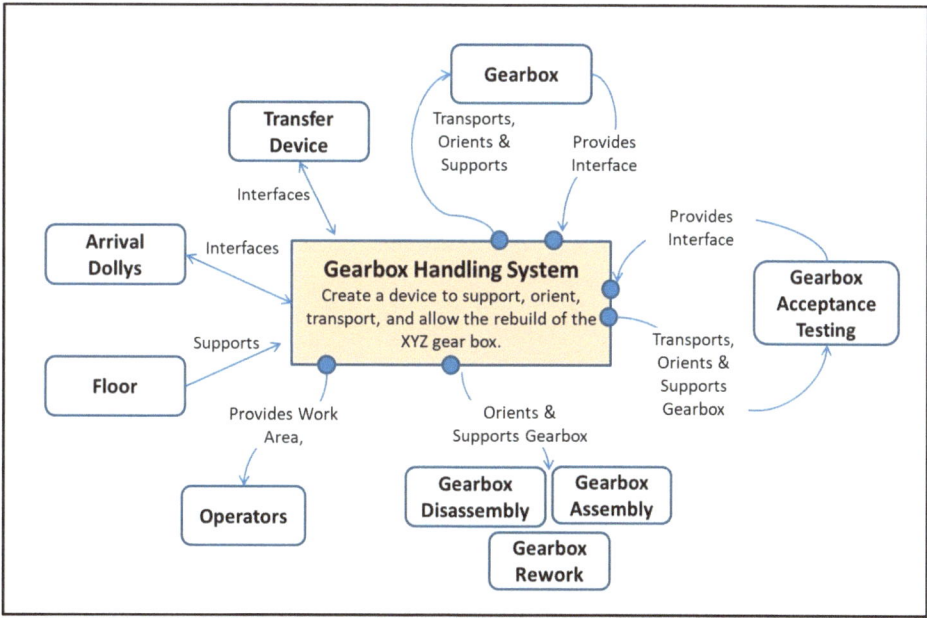

Figure 67: Gearbox Handling System - Operational View Context Diagram

You also develop the Detailed Functions Outline for your system (Figure 68) which shows WHAT your system must do to satisfy the customer's NEED.

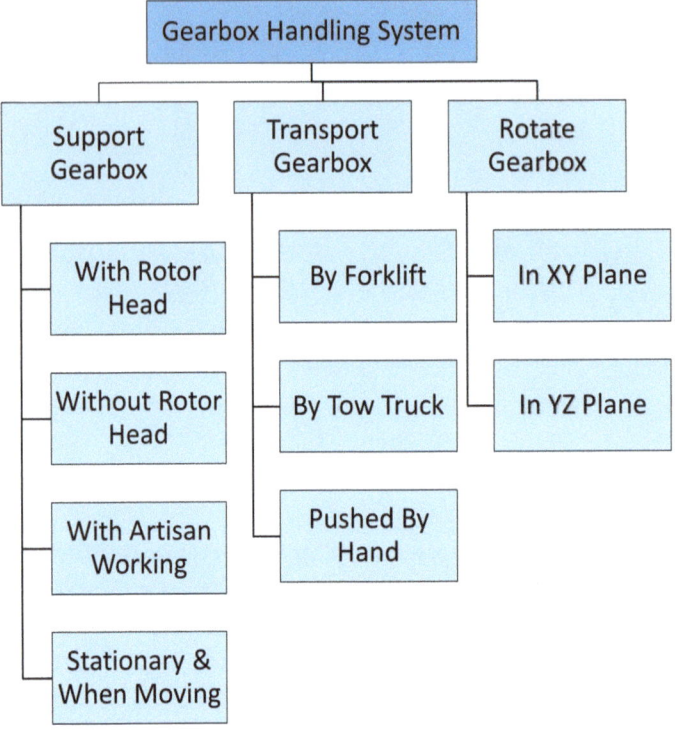

Figure 68: Gearbox Handling System - Detailed Functions

You continue exploring needs until you have discovered everything necessary to satisfy the customer's NEED. You know more details may surface as your design develops (remember the ship approaching shore analogy on page 32), but you have uncovered everything you can at this point. A few examples of the needs you discover are as follows:

1. The device needs to satisfy the orientation requirements shown in Figure 63. (Requirements 3, 4, 5)

2. The device serves as the work stand during the rebuild, so your device needs to allow the operator to stand on it and perform work on the gearbox. (Requirement 6)

3. The device will be pulled throughout the plant by a tow truck. (Requirements 2, 7, 8)

4. The device is sometimes lifted by a forklift. (Requirements 2 and 9)

5. The device needs to allow access to the area behind the mounting plate so it can stay on the device during Component Clean and Component Paint. (Requirement 10)

Once you have discovered all of the needs (remember, needs and wants), you work with the customer and decide exactly which needs will be addressed by your design. Where necessary, operational needs are further defined, and then all needs to be addressed by your system are collected to form the system requirements. Some of the requirements for the example needs above can be seen in Figure 69. By each example need above, the corresponding requirements in Figure 69 are shown. Needs 1 and 3 with their corresponding requirements are examples of operational needs that required further definition for design requirements.

You formally present the requirements to the customer to receive their approval. Together, you agree on the requirements, how you will prove that your system performs as promised, and the criteria for final acceptance.

---

**Gearbox Handling System Requirements**

**Must Have**
1. The device shall interface with the gearbox according to drawing AB123.
2. The device shall support the gearbox in all states (static, manual movement, fork lift movement, tow truck movement).
3. The device shall have a +90° rotating range in the XY plane.
4. The device shall have a -25° rotating range in the XY plane.
5. The device shall have a -25° rotating range in the XZ plane.
6. The device shall allow the operator to tear down and rebuild the gearbox while the gearbox is mounted to the device.
7. The device shall be able to be pulled by a tow truck.
8. The device shall contain a 3" receiver to interface with a 3" ball mount on the tow truck.
9. The device shall be able to be picked up and transported by an 8,000 pound capacity fork truck.
10. The entire back of the gearbox (within 4" of the mounting bolts) shall be accessible for cleaning and painting of the gearbox.

**Like to Have**
1. The device should allow the gearbox to be transported by one operator.

---

Figure 69: Gearbox Handling System - Requirements for Example Needs

**Conceptual Design**

You enter Conceptual Design with fear and trembling and for good reasons. Consider the following.

- Your work during Conceptual Design will set the direction for the rest of the design project. The decisions you make during Conceptual Design have far-reaching consequences and will largely define the design's performance potential, reliability, cost, and whether or not the customer's NEED is satisfied.

- You have a set of requirements, but that is all. You start with nothing and end with the full concept for a system to satisfy the customer's NEED. This is a daunting task!

- You enter Conceptual Design with a solution already in your head, and you find it hard to think about anything else but this solution. You remember the importance of not jumping to a solution too early (page 6), but it is very difficult to practice the Discipline of WHAT before HOW and follow the Design Thinking Cycle (page 51) to generate fresh ideas.

- There is no "formula" that you can apply to every problem and arrive at a conceptual design. Of all the stages of design, Conceptual Design is the one with the most iteration and bouncing back and forth between high-level and low-level details.

Even though you are slightly intimidated, you start Conceptual Design by gathering and reviewing the information below.

- Needs Statement (Figure 65)
- Wider System Diagram (Figure 66)
- Context Diagram (Figure 67)
- Functions Outline (Figure 68)
- Requirements (Figure 69)

Together, these documents give you a clear picture of WHAT your system must accomplish and allow you to visualize the wider system easily. You will research and explore many ideas during conceptual design, and you should review these documents often to keep your focus on the unique aspects of the complex problem situation and on what is necessary to satisfy the customer's NEED.

For this design, you choose to start from the bottom up, so you research and brainstorm ways to accomplish each top-level function in Figure 68. For example, you research and brainstorm ways to rotate the gearbox in the two different planes. While focusing on the functions, you also brainstorm the system as a whole. What you learn when focusing on the functions influences your whole systems brainstorming, and your whole systems brainstorming causes you to go back and consider other approaches to accomplish the functions.

You refine your design by continuing to bounce back and forth between the big picture and details following the Design Thinking Cycle from page 51. To ensure you select alternatives like the customer would, you use the Evaluation Criteria when evaluating alternatives. You often use your technical skills to perform high-level analysis on an idea to ensure it is feasible. For example, you have the idea of supporting and rotating the gearbox in the YZ direction using two pins. You perform a rough strength analysis calculation to get an idea of how large the pins must be. Reasonably sized pins can support the gearbox, so you explore the idea further. This process continues, and eventually, you develop the conceptual design shown in Figure 70.

Your conceptual design does not contain all of the details of your final design, but you do show the customer that your conceptual design 1) is feasible and 2) has everything required to satisfy their NEED. Notice that images 1, 2, and 3 show the orientation capability for the X-Y plane, and that image 4 shows it for the Y-Z plane. Images 5 and 6 show how you will accommodate the X-Y rotation while maintaining access to the center area.

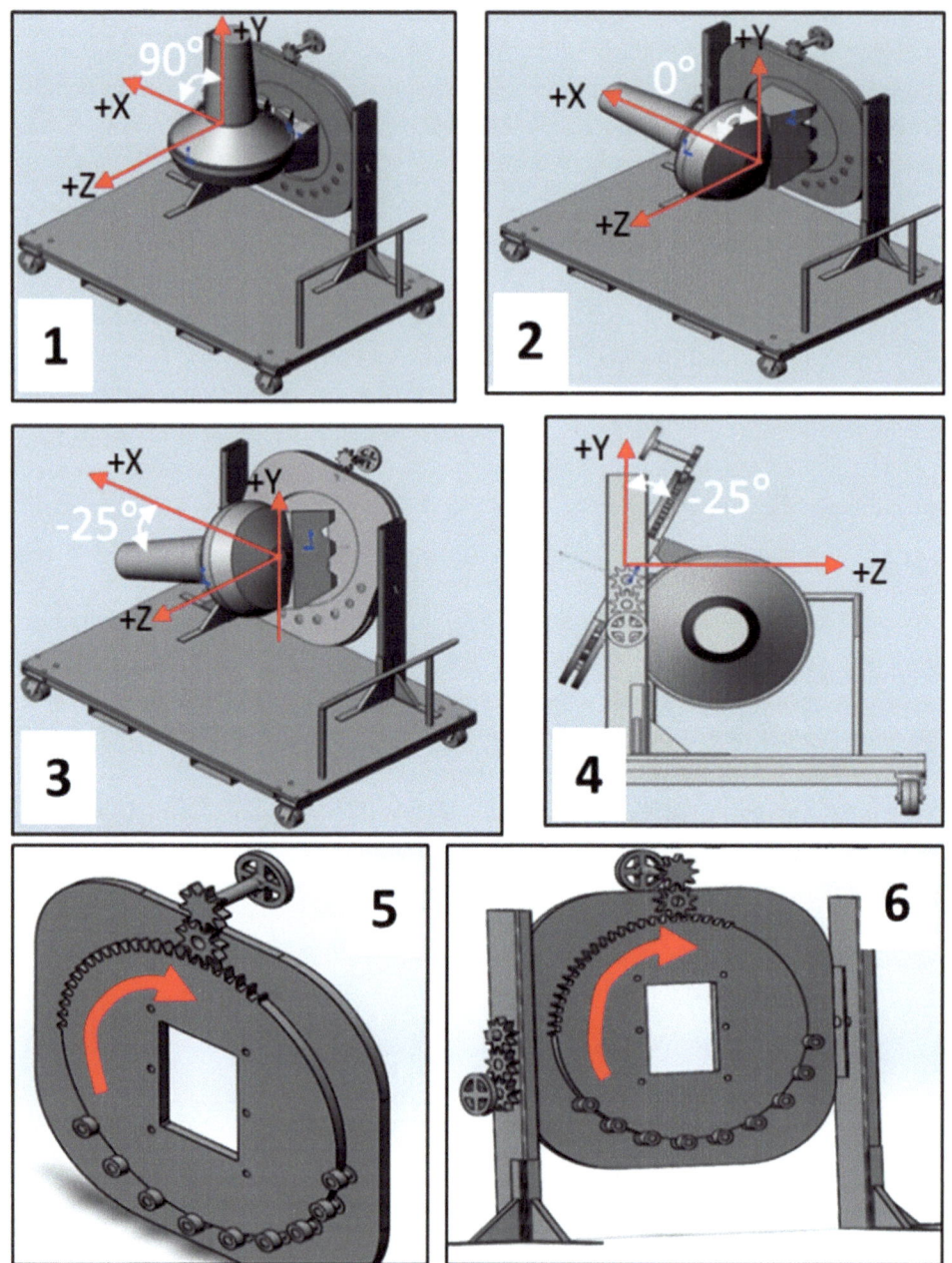

Figure 70: Gearbox Handling System - Conceptual Design

## Preliminary / Detailed Design

Based on the requirements and your conceptual design, you perform preliminary design and then detailed design to engineer a complete solution.

At the end of conceptual design, you split the overall system down into the sub-systems shown in Figure 71 (review Figure 25 on page 44 for a reminder of how sub-systems fit into the stages of design). A team member is given responsibility for the design of each sub-system, and a team member is also assigned the integration responsibility for the full system. The integration team works with the sub-system designers to ensure the details of each sub-system are integrated for the best overall system performance.

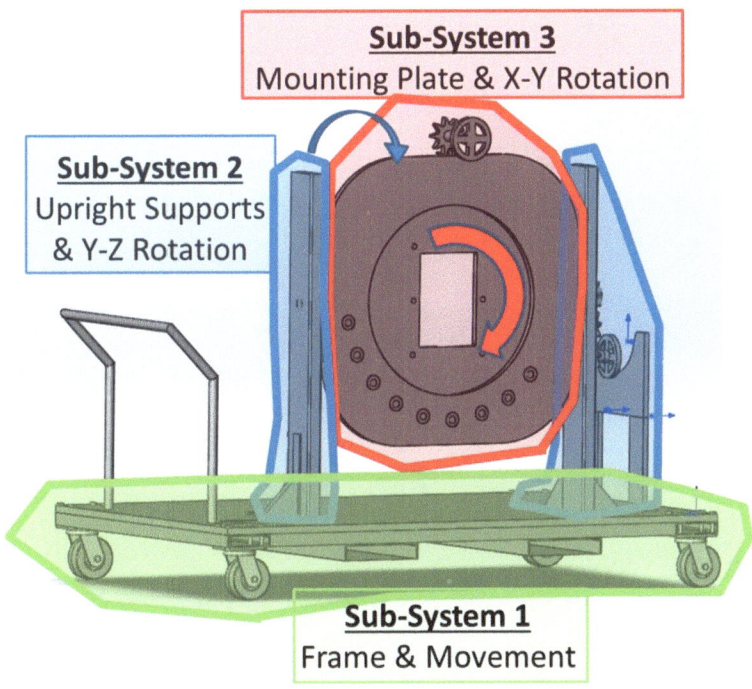

Figure 71: Gearbox Handling System - Sub-Systems

Each sub-system designer defines requirements for their sub-system and then uses Design Thinking to develop a sub-system that satisfies all requirements. While designing a sub-system, you will have the sub-system requirements to guide you, but you will not just rely on these requirements. As you work through the details of your design, it is possible that details are emerging now that may not be covered by the requirements. You did the best you could defining requirements, but now that you understand the problem even better, there may be some things that "matter" to the customer that were missed (remember the ship approaching shore analogy).

Periodically, you look back at your purpose (Figure 65 Needs Statement) and the entire system (diagrams in Figure 66, Figure 67, and Figure 68). As new details emerge, you work with everyone impacted and adapt.

For an example of details that emerge, consider what happens when you are performing strength analysis on the elements of your sub-system. You remember that the dolly serves as the work stand during the rebuild and that it will be pulled throughout the plant by a tow truck. You have included interfaces for the tow truck in your design as well as made the platform large enough for the operator to stand and work. What you have not done is taken into account the loads that these activities will place on your device. The operator will put force on the gearbox when working and this force will be translated to the device. If the tow truck stops abruptly, the gearbox will have momentum that will be translated as a force acting on your device. Now that you are closer to shore and details are coming into view, you realize the existence of these forces.

The systems perspective helps you realize the existence of the forces, and now you use your traditional engineering design skills to calculate the values for all of the forces acting on the gearbox. The result is shown in Figure 72.

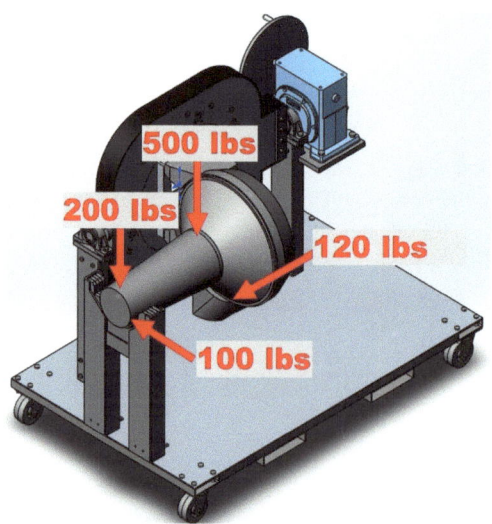

Figure 72: Gearbox Handling System - External Forces

With help from the integration team, you share the information from Figure 72 with all other sub-systems so everyone will accurately perform strength analysis on all elements within all sub-systems.

Once again, notice the systems perspective working with your traditional engineering skills. The systems perspective allows you to identify what is required for the design to satisfy the customer's NEED. Your traditional analytical engineering skills allow you to design a device to accomplish what is needed.

During detailed design, you are immersed in defining every aspect of your design. You use your analytical tools to ensure each component is designed to perform its function safely. For example, consider Figure 73.

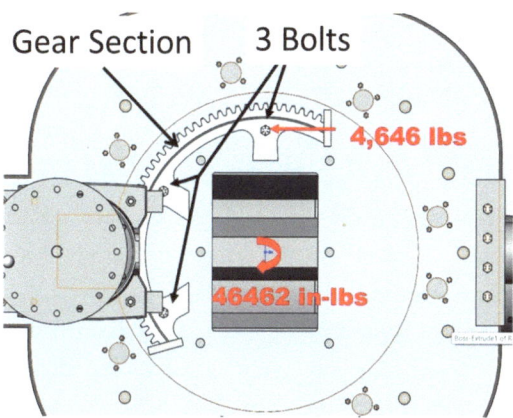

Figure 73: Gearbox Handling System - Example Strength Calculation

The gear section that rotates the gearbox is attached using three bolts. The bolts experience specific loads during use, so you must perform strength analysis to determine the bolt size required for safe operation. When sizing your bolt, you consider what is commercially available, so time will be spent researching to determine viable options. When done, every detail of the bolt will be specified so that it can be purchased (type, grade, length, diameter, thread). The specification of these details is an example of Component Requirements shown in Figure 25 (page 44) and Figure 29 (page 50). Notice in both figures how requirements flow all the way from the customer's NEED to component requirements. Specifying the details for a component ensures that it properly performs its role and satisfies the sub-system requirements. Satisfying the sub-system requirements ensures that each sub-system performs its role in the overall system and satisfies the system level requirements. Satisfying the system level requirements allows your system to satisfy the customer's original NEED.

The level of detail explained in the example above is repeated for every aspect of your design. Every bearing, gear, purchased part, and manufactured part is designed and analyzed to ensure it is safe and that it satisfies all requirements.

While working on the details of your design, it is very easy to lose sight of everything but the details. For example, all you can think about is how to rotate the 500-pound gearbox without a center rotation point. Now and then, you stop and look up from the details. You review the big picture documents created when defining requirements.

- Needs Statement - Figure 65
- Wider System Diagram - Figure 66
- Context Diagram - Figure 67
- Detailed Functions - Figure 68

You look around and remember that you are in a system. You remember all of the points made about a system on page 21 and that your sub-system is only a part of the full system. You remember the discipline of WHAT before HOW explained on page 7. You remember the "4 Cardinal Rules for an Effective Design" on page 54. You remember to consider everyone else in the system and to make sure your details fit in with all of the other parts. You remember to ask "what do the other elements want out of me?" and then make sure you give it to them. You also remember to ask "how do the other elements influence me?" In all of this, you remember that you must consider the wider system as well as the other sub-systems that make up your design. You make sure you are communicating with all these other elements and that you take the interactions with them into account with your design.

When all details for each sub-system have been defined, the Integration team helps you and the other sub-system design leaders bring each sub-system together into the final design package. You then compile all of the details needed for your design to be built. At a minimum, you will have

- a complete list of purchased parts,
- assembly drawings,
- detailed engineering drawings for all manufactured parts,
- and technical analysis supporting all critical design decisions.

You present all of this information along with your technical analysis at one final design review. After addressing any issues discovered during the design review, your design is complete and ready to be built.

An example of the final solution for the gearbox example is shown in Figure 74

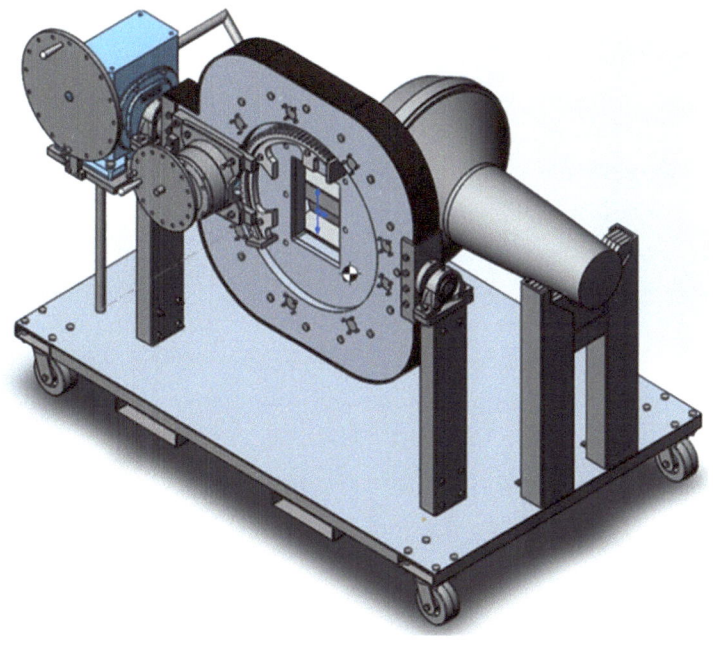
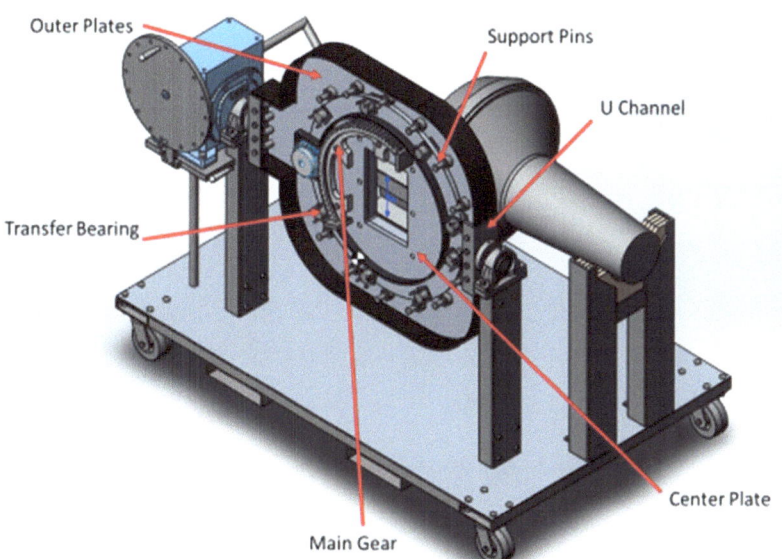

Figure 74: Gearbox Handling System - Final Design

**Summary**

This chapter allowed you to see how the systems perspective works with your traditional engineering design skills to satisfy a customer's NEED. You saw the systems perspective complementing your traditional technical skills in a way that allowed you to provide a solution that saved time, saved money, and reduced the potential of injury to people and damage to the gearbox. As is often the case, the customer in the Gearbox Rebuild System only understood the NEED from their perspective. Systems are always composed of many perspectives, and the customer needed you to help them see the wider system and how adding an "unnecessary" requirement to "their" device would have a significant positive impact on the entire system. Together, the systems perspective and your traditional technical design skills allowed you to satisfy the customer's true NEED and leave behind a very satisfied customer. You could not have done this using either set of skills alone.

# Chapter 6 – Performing on a Design Team

*A team is where a group of people learn how to utilize the unique strengths of each person on the team to accomplish a specific task better than any single person on the team could complete it alone.*

---

Functioning as a team is an extremely difficult task. It is easy to divide work up among people to accomplish assigned tasks, but that is not functioning as a team. A team is where a group of people learn how to utilize the unique strengths of each person on the team to accomplish a specific task better than any single person on the team could complete it alone. Being a true team takes work, and you must deliberately take steps to function as a team. I will talk later about some helpful roles for your team, but first, consider two essential parts of every team – team members and a team leader. Opening remarks to each group are given below.

For an effective team, you as a team member must decide to do the following:

- Believe that every member of your team has something to offer and that the task at hand will get completed better because they are there.

- Commit to learning something from each one of your teammates.

- Remember that you are not the only one with great ideas, so commit to truly listening to the thoughts and ideas of your teammates.

- Listen to and watch your teammates and understand their strengths (every member has some).

- When you get your feelings hurt, which you will, take time to cool off and then effectively communicate with your team about the issue. Forget about your feelings and about "winning." Seek to come up with the best idea for the task at hand.

- Do not let the team down by doing sloppy or incomplete work.

- Do not rely on the team leader to complete the task at hand. Feel responsible for your team completing the task with excellence.

You as the team leader must understand the importance of your role.

- People often think of the leader as "the person in charge." The leader is the one serving the rest of the group by looking ahead to identify what needs to be done and then helping the rest of the group to engage in the required tasks.

- The group leader is the one that has agreed to serve the team by doing the extra work of helping everyone else know what they need to be doing.

- Practically speaking, the leader is typically the person most engaged in the task at hand. Often in business, the leader is the only person engaged in looking at the full task at hand. The other members are doing their part and are counting on the leader to tell them when they need to pay attention to something. If they do not hear anything from the leader, they assume all is well.

- Your role as a leader is critical because you are ultimately responsible for getting the task at hand done and for managing group dynamics to help the group function as a team.

- If the project is a success, the members of the team get the credit. If the project fails, the leader takes the blame. The leader does not have to do everything, but they are the one responsible for making sure everything is done.

Please take this opportunity to improve your ability to work as a true team. It is an investment that will give dividends for years to come! Below are some tips to help you perform as an effective team.

**Attitude is Everything**

OK, let's be honest. It is likely that you think teams are kind of a pain. You may even think that teams are a waste. I know, you can do the task just fine by yourself, and everyone else on your team just slows you down. Am I close? I have been there, but let me assure you that you need your teammates. You need them to do a better job than you can do alone. Please accept this fact and adopt the attitude that your teammates are essential to good performance. Yes, I know it is hard.

No matter how smart someone is, they have blind spots. Yes, even you have blind spots. The people on your team help fill in your blind spots, so together you see much more clearly and completely. Together you can do a better job.

No matter how capable a person is, they have weaknesses. Your teammates will have strengths that you do not have. If your team learns how to utilize the strengths that each person brings, you can accomplish things that you would never attempt alone.

Attitude is everything. See the other people on your team? You need them. They see things you do not. They have strengths that you do not have. They are a key part of your success. Together, you can do great things.

**Define Team Roles**

Having some formal team roles is important, and three key roles are described below. It is important to assign these roles specifically. Usually, when no one is responsible for something, it will not get done.

Recorder – During meetings, the recorder listens for and documents **decisions made, action items**, and **points that need to be clarified**. These items need to be documented in meeting minutes. Often your success on a project hinges on how well you can identify and capture key details in a meeting. If a customer communicates a requirement in a meeting and you miss it, you are in for big trouble. Sometimes you have to hear what is not being said! The customer may make a side comment that reveals an aspect of the system that has not been discussed, and you must catch this detail. The recorder should be actively listening to identify these small details. The team leader often will hear a specific detail and ask the recorder to make sure they get it down.

Facilitator – The facilitator is the one at group meetings that focuses on and helps the group to function as a team. The facilitator monitors the meeting progress to keep it on the subject and on time. Groups can often bounce around from topic to topic and never get anything accomplished. The facilitator helps the group to focus on one thing until a decision is made or an action is assigned. If important topics come up while discussing something else, they should be put in a "parking lot" to be returned to later. The facilitator also is the one that ensures everyone is participating. Sometimes this means to ask a member for their thoughts, and sometimes it means asking a member to give someone else a chance. Healthy conflict can be very productive in a group, and the facilitator is the one that helps keep conflict positive when it arises.

Team Leader - People often think of the leader as "the person in charge." The leader is the one serving the rest of the group by looking ahead to identify what needs to be done and then helping the rest of the group to engage in the required tasks. The group leader is the one that has agreed to do this extra work and help everyone else know what they need to be doing.

As the leader, you need always to be looking ahead and see what you need to be doing. You need to be thinking about the big picture and the entire task and then planning the small pieces that need to be done right now to get there. You are the project manager.

All of your planning work usually comes out in meeting agendas. You are the one that should be pulling the group together for face to face at meetings. At meetings, you should have thought through what needs to be done and then lead the group in doing it. At meetings, you will also make sure new tasks are assigned when it comes time to start them.

**Resolving Conflict**

You are going to have conflict in your group. All teams have conflict, but teams of engineers, definitely have conflict. Disagreeing and working through ideas is part of the engineering process. When you and another team members both have an idea of what is best, what do you do? An effective team does the following:

1) Each idea is picked apart (that means you get grilled, and there is plenty of opportunity for conflict).
2) The team uses facts to determine what will and what will not work (your perfect idea just went down the tubes, so more chances for conflict).
3) The team takes the best aspects of all ideas and comes up with the ideal solution to the problem (nothing from your idea was included, so once again more opportunity for conflict).

The three steps above describe the reality of engineering. You are going to share an idea, and a peer engineer will challenge the idea. At first, you will not like it, but you do want someone to challenge your ideas and find any holes. You want your peers to find the issues with your ideas instead of you discovering them after you build your design. Challenging an idea is a normal part of group activity, so you want to learn to do it positively. Below are some tips to help when you challenge an idea or have your idea challenged. Giving and accepting criticism is a tough area, so be ready for some deliberate practice to become proficient at it.

<u>Learn not to take it personally when someone challenges your idea</u>
When someone challenges our idea, we can respond in one of two ways.

> Response 1: We can realize that the challenge is a normal part of the engineering process and have an adult conversation with the person to arrive at the best solution. We will hear their criticism and seek to offer a logical defense when appropriate. We will have a healthy dialogue. When they uncover a true flaw in our idea, we will acknowledge it and gain insight by talking through the issues with them. In the end, either we will abandon the idea, or we will take their insights and improve it.

Response 2: We can take the challenge to our idea personally, and then feel the need to defend ourselves and show that we are right. In doing this, it is likely that we will get angry. Our anger will come out in one of two ways. Our anger may cause us to dig in even harder with our attempt to show how we are right and that the other person is wrong. When this happens, all logic is out the window and the problem at hand is secondary to winning. Emotions are likely to escalate. The other way our anger can come out is by withdrawing. We get our feelings hurt and check out. We take our toys and go home. The result is the same whichever way we respond. We hurt the effectiveness of our team, and we are no closer to finding the best solution than when we started. The customer is the loser in this situation.

<u>Always remember that the goal is to find the best idea. It is not to win!</u>
- As an engineer, you like to be right. We all do and that is okay, but you have to realize that you are not going to be right all the time. You are in school with a lot of other smart people, and they will be right sometimes as well.
- It is never about winning. It is always about finding the best solution.

<u>Actively listen and fully understand the other person's point.</u>
- Truly actively listen to the other person to understand their point. The other person is smart just like you, and you want to make sure you understand their insight.
- Do not formulate your next point while the person is still talking. Formulating your point while someone is talking is a "winning" mentality, and you should avoid doing it.
- To make sure you understand the other person's point, ask clarifying questions or summarize what you think they are saying.
- Acknowledge when the person has a valid criticism of your idea. Hear the good points they make and incorporate their insights into your idea.

<u>Stick to the facts</u>
- Engineers make decisions on facts, not on opinions.
- Separate the facts from opinions. It is easy to say, "I do not think that will work," but that is an opinion. When someone gives an opinion, ask questions to determine the facts. For example, "What about it do you think will not work?" or "Why will it not work?"

<u>Make comments in a constructive way</u>
- Think before you speak. I think we all know what this means.
- When possible, phrase your feedback in a positive tone. For example, say something like, "Could you explain how that would work?" instead of "That does not make any sense."

- Criticize ideas and not the person. Realize that the other person is not criticizing you. They are criticizing your idea, and you need to separate the two.
- Don't just find problems. It is easy always to be negative and find problems with ideas. Along with identifying problems, try and find a solution to your objection.

<u>Know when to take a break</u>

At times, group discussions can get a little intense, and you may feel your emotions rising. When this happens, the best thing to do is to take a break. Tell your group that the discussion is getting a little intense for you and that you are going to take a break. Walk away, calm down, and then come back ready to contribute. Walking away is a lot better than having to apologize for saying something that you really did not mean.

**Summary**

As an engineer, working in a team will be part of your life. Use the tips above, and learn how to be an effective member of a high performing team.

# Chapter 7 - Before You Begin Your Design Project

*The customer needs your help, and they are placing great trust in you by coming to you to satisfy their NEED.*

---

This book has given you the mindset and tools you need to get started towards a successful design project. You can define requirements for your problem, and you have some new thinking that will help you during all phases of design. Before we end our journey, I give you a few personal lessons learned during my design projects.

**Be Patient**

Be aware that applying this material to your design problem will take some time to learn. This is normal, so do not get discouraged and quit. You are learning new skills, and it is going to take some practice. The time invested in learning these steps will yield great dividends. Once you are comfortable with the steps to define requirements, you will see that using them for most problems takes very little "extra" time. Depending on the complexity of the problem you are solving, you may go through the steps informally in your head or in a systematic manner with formal documentation. Either way, the thinking embedded within each step will lead you to explore everything necessary to define the problem in a way that will allow you to create a design that satisfies your customer's NEED.

**Focus on WHAT and Resist Thinking about HOW**

I discussed this when the discipline of WHAT before HOW was presented, but it is important enough to say again. When defining needs, you must resist the natural pull of your mind to jump from WHAT to HOW. As soon as your thinking goes from WHAT to HOW, you limit yourself. When you move from thinking about WHAT to thinking about HOW, you move from considering all possibilities to focusing on just one. Most design breakthroughs come while thinking creatively about WHAT.

When you define a HOW as a need, you place design restrictions on yourself and limit possible solutions. You only want to do this when it is necessary. If the customer tells you that all communications in their facility must be done using Bluetooth, then this is a HOW that you must define as a constraint. Constraints are the only time you want to define a HOW as a need, but do not do so without asking questions to ensure it is necessary. Do not just accept constraints without probing to make sure they are real. Constraints limit your design choices so only include a constraint when you are convinced that it is absolutely necessary.

**Manage Your Personal Interactions**

Since most of us engineers are not always that great with people, a few words of caution are in order. When defining requirements, you will be getting much of your information from people, and people do not always completely understand the situation or know exactly what they want. Your ability to ask respectful, insightful, and clarifying questions is the key to your success! In time, it is likely that you will understand the problem and the situation better than the customer. Because of this understanding, your job is often to listen to and then educate the customer on what is necessary to satisfy their perceived NEED. The idea that "the customer is always right" does not really hold in systems development. You always need to make the customer feel heard and respected, but sometimes satisfying their true need is helping them understand that they are asking for the wrong thing.

**Learn to Listen**

Never underestimate the importance of listening. Often the customer will not just come out and tell you about a need, because the true need may not be known. You have to learn to listen carefully and evaluate how what they say relates to what they are asking you to do. You have to learn to "listen between the lines" with insight to evaluate how the design project is impacted. By careful listening, you can often identify needs that end up being critical pieces to the success of your system. Listening in this way takes practice and is a skill you should develop.

**Explore Needs as Closely to the User Level as Possible**

When defining needs, you should remember that the "*customer*" is not just the one that asked you to do the work. The customer you are concerned with at this point is everyone in the wider system that influences or is influenced by your system. It is always best to define needs by getting as close to the user level as you are allowed to go. It is easy for those in offices to forget the details of the problem situation, so don't just ask what is important to a group like the operators or maintenance. Go and talk with them yourself.

As you talk with various groups, you will find yourself working with people that are lower than you on the organizational chart. Their position on the organizational chart does not reflect their ability to contribute to defining needs for your system. People that work the closest to the system (operators, maintenance, shop floor workers) have insights into the system that you and others in the office will never have. They are a wealth of knowledge, and they deserve your utmost respect. Give them this respect, and they will gladly help you understand the true system needs. You also may find yourself somewhat intimidated when having to talk with people in the production environment. This is normal. You are an engineer, and you like to feel competent, but you may not feel this way in the production environment. Do not compensate for your insecurity with a false sense of confidence that will come across as arrogance. Instead, respectfully ask questions, actively listen, and learn. Pairing your technical knowledge with the intimate practical knowledge from those close to the system is how you create great systems.

**Know When to Let Go**

We all have our pet ideas that we become emotionally attached to. The idea is like our child, and we fight to make it work. Working hard to overcome obstacles with an idea is a good thing. You do not want to abandon an idea just because it has obstacles to overcome. However, you must know when enough is enough. You have to recognize when you are trying to force your pet idea to work when in reality it is not a fit for the problem. Even if your idea will work, it still may be too complex, too costly, or take too much time to develop for the current problem situation. When it is not a fit, you have to let it go. If you do not let it go, it will pull you down like holding on to a boat anchor in the water.

**Know When to Quit**

There is general rule used in many areas known as the law of the vital few or the 80/20 rule. It is not a scientific law, but it is a principle that holds in many cases that 80% of the effect will be achieved by taking care of 20% of the causes. This principle is illustrated for design in Figure 75.

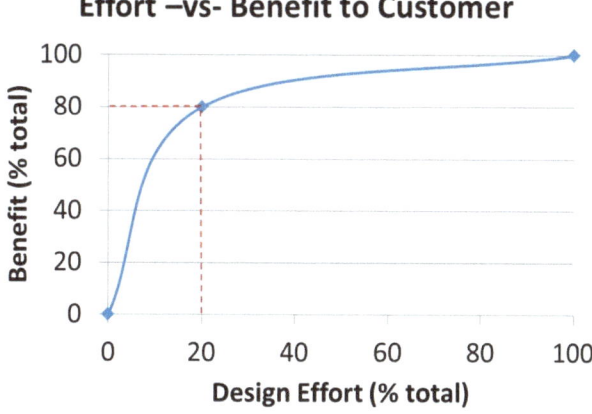

Figure 75: 80/20 Rule for Design

The numbers are not exact, but the idea is that a small amount of the design effort will satisfy most of the customer's NEED. Satisfying the last 20% of the NEED will require significantly more effort, and the effort may not be worth the benefit. Effort means cost and time to complete the project, and you will always have a limited amount of both. You can spend a lot of time, money, and energy trying to satisfy a few small needs and end up not satisfying any needs. In trying to satisfy every little need, you can cause the entire project to be a failure by creating a system that is too complex, too expensive, or takes too long to develop. Successful designers know how to work with the customer to determine when enough is enough. As an engineer, you will want to make everything "the best" it can be, but learning when to quit is a critical skill you must develop.

**Closing**

This book opened with you being asked to design something, and I told you that we were going to go on a journey together. Since you are reading, I assume you persevered and completed the journey. Thank you for continuing to the end.

It is now your time to go and perform engineering design. This book and your engineering courses have prepared you, so go in confidence knowing that you can be successful. I end our journey together with a list of all the reasons I think you will be successful in your design adventure.

- You understand your role as a "Needs Satisfier" (Figure 11 on page 16).

- You understand the Discipline of WHAT before HOW (Figure 3 on page 7).

- You understand systems (Figure 17 on page 26) and the systems perspective (Figure 18 on page 27).

- You understand that properly defining requirements is essential to a good design, and you know how to develop requirements (Figure 32 on page 56).

- You have the stages of design to lead you through your design (Figure 25 on page 44 and Figure 29 on page 50), and you understand WHAT must be accomplished at each stage for a successful design.

- You understand design thinking (Figure 30 on page 51).

- You know the 4 Cardinal Rules for an Effective Design (Figure 31 on page 54).

- You have sound technical knowledge from your engineering courses.

The only thing left for you to do is to go and satisfy your customer's NEED through your design. The customer needs your help, and they are placing great trust in you by coming to you to satisfy their NEED. Like Bob, I know you will not let them down.

# About the Author

Dr. Fortney was trained in advanced manufacturing at Purdue University and received his Ph.D. in Industrial Engineering from the University of Tennessee. At UT, Fortney's research focused on systems approaches to organizational improvement, and after graduation, he utilized the systems perspective to design and install automated manufacturing systems for MasterBrand Cabinets. Fortney's work at MasterBrand's Crossville, TN plant resulted in a system called "the rough mill of the future" and was featured in the November 1996 issue of Wood & Wood Products. His other roles at MasterBrand include corporate industrial engineering manager and general manager of the Kinston, NC Facility. As general manager, Fortney used the systems approach to design, create and operate all aspects (human resources, production control, manufacturing) of a new 600,000 $Ft^2$ cabinet facility.

Since MasterBrand, Fortney has been with North Carolina State University developing a new site-based engineering program focused on the unique needs of the Fleet Readiness Center East Research and Engineering Group for the Naval Air Systems Command. His ABET Accredited Mechanical Engineering Systems (MES) BSE program now provides NAVAIR with a steady stream of highly qualified mechanical engineers for their Havelock, NC facility. While creating the MES program, Fortney has focused on developing techniques to teach the systems perspective for design to undergraduate mechanical engineering students. He has embedded these techniques within his MES senior capstone design course as well as within a junior level project-based course where students learn and experience the System Engineering approach.

Outside of work, Fortney enjoys serving at church with his wife and playing with his many grandchildren.